AF454171

AGRICULTURAL
SUPPLY CHAINS
IN NIGERIA

From Farm to Market Efficiency

Jumoke Ayodele Raji-Ayoola

Published in Nigeria in 2024 by Emphaloz Publishing House
under its Emphaloz imprint

www.emphaloz.com

A catalogue record of this book will be available from the
National Library of Nigeria.

TABLE OF CONTENTS

PREFACE

Agriculture lies at the heart of Nigeria's economy, culture, and history. It sustains millions of livelihoods, feeds the nation, and holds immense potential for driving economic growth and global competitiveness. As one of the country's most significant economic sectors, agriculture is uniquely positioned to address pressing challenges such as poverty, unemployment, and food insecurity. However, despite its potential, the sector faces numerous challenges, including inefficiencies in supply chains, inadequate infrastructure, limited access to markets, and systemic barriers that hinder innovation and growth. These issues not only curtail productivity but also perpetuate cycles of poverty, food shortages, and economic stagnation in rural areas.

This book, Agricultural Supply Chains in Nigeria: From Farm to Market Efficiency, is a comprehensive exploration of the opportunities and solutions that can transform Nigeria's agricultural sector into a thriving engine of development. The book takes a holistic approach, examining the interconnected components of agricultural supply chains; production, processing, storage, transportation, and marketing while identifying systemic bottlenecks that undermine efficiency. Through an actionable framework, it proposes strategies to address these challenges, emphasizing innovation, collaboration, and sustainability as drivers of transformation.

Drawing from extensive research, case studies, and practical insights, the chapters in this book delve into key topics such as technology integration, infrastructure development, financing mechanisms, sustainability practices, and policy reform. Agricultural builds upon the others to provide a comprehensive understanding of how Nigeria's agricultural supply chains can be reimagined and optimized to deliver better outcomes for farmers, businesses, and consumers.

This book is written for a diverse audience, including policymakers, agribusiness leaders, farmers, researchers, and development practitioners. Whether you are an academic studying agricultural system, a policymaker crafting strategy for rural development, or a farmer seeking to navigate market dynamics, this book offers valuable insights and practical solutions tailored to Nigeria's unique context. By bridging knowledge gaps and highlighting successful examples, it provides a roadmap for achieving efficiency, equity, and sustainability in the agricultural sector.

As you journey through these pages, I invite you to reflect on the immense potential of Nigeria's agriculture and the collective effort required to realize it. The path to transformation is challenging, but the rewards are profound, a resilient economy, improved livelihoods, and a sustainable future for generations to come. This book is not just a guide; it is a call to action for all stakeholders to reimagine agriculture as a cornerstone of Nigeria's development.

FOREWORD

A griculture is more than an economic activity; it is a lifeline for communities, a source of nourishment, and a pillar of national identity. In Nigeria, agriculture employs over 70% of the population and contributes significantly to the GDP. Yet, the sector's potential remains largely untapped due to inefficiencies in supply chains, outdated practices, and structural challenges that hinder growth. Addressing these issues is critical for achieving food security, reducing poverty, and fostering economic resilience.

This book, *Agricultural Supply Chains in Nigeria: From Farm to Market Efficiency,* arrives at a critical juncture. As the world grapples with the dual challenges of food insecurity and climate change, transforming agricultural supply chains is no longer optional, it is imperative. With its vast arable land, diverse agro-ecological zones, and entrepreneurial spirit, Nigeria is uniquely positioned to lead this transformation and establish itself as a leader in sustainable agriculture.

What sets this book apart is its holistic approach. It examines every link in the supply chain, from farm to market, and provides a clear roadmap for addressing inefficiencies. By focusing on infrastructure development, technology adoption, financing mechanisms, and sustainability, the book outlines practical strategies for overcoming systemic barriers. Its emphasis on inclusivity ensures that marginalized groups, including smallholder farmers, women, and youth, are empowered to participate fully in the agricultural transformation. Furthermore, the book aligns with global trends, integrating principles of sustainability and innovation to position Nigeria as a competitive player in international markets.

I commend the author for crafting a work that is both visionary and practical. This book is an invaluable resource for anyone seeking to understand and improve Nigeria's agricultural sector. It challenges us to think boldly, act decisively, and collaborate meaningfully to unlock the full potential of agriculture. The insights within these pages inspire optimism and a renewed commitment to ensuring that agriculture remains a cornerstone of Nigeria's development. I am confident that it will inspire readers to join the movement for a brighter, more sustainable future for Nigeria and beyond.

INTRODUCTION

Agricultural supply chains are the backbone of any food system, connecting producers to consumers and transforming raw commodities into value-added products. In Nigeria, these supply chains play a pivotal role in ensuring food security, driving economic growth, and supporting millions of livelihoods. Yet, they are fraught with inefficiencies that weaken their impact. Post-harvest losses, inadequate storage, poor transportation networks, and limited market access are just a few of the challenges that undermine productivity and resilience. These challenges disproportionately affect smallholder farmers, who make up the majority of Nigeria's agricultural workforce and often face systemic barriers to growth.

This book begins with a fundamental question: *How can Nigeria's agricultural supply chains be transformed to achieve efficiency, sustainability, and inclusivity?* The answer lies in addressing systemic issues and embracing innovative solutions tailored to Nigeria's agricultural landscape. Through a comprehensive exploration of key themes, the book seeks to provide actionable insights that guide stakeholders toward meaningful change.

Each chapter examines a critical aspect of supply chain transformation, beginning with an analysis of the current state and progressing toward practical strategies for improvement. The book explores the role of infrastructure, logistics, and technology in optimizing supply chains;

examines financing mechanisms that empower stakeholders; and highlights the importance of sustainability in ensuring long-term viability. It also emphasizes policy reform, capacity building, and collaboration among stakeholders as essential pillars for driving change.

The insights and recommendations presented here are informed by a blend of research, case studies, and success stories from Nigeria and beyond. By integrating these diverse perspectives, the book offers a roadmap for addressing inefficiencies while fostering innovation and resilience. It aims to equip readers with the tools to understand and address the complex challenges of agricultural supply chains in Nigeria.

Agriculture is not just a sector, it is a foundation for development, resilience, and prosperity. Transforming Nigeria's agricultural supply chains requires a collective effort involving farmers, policymakers, private sector actors, and development partners. This book is a call to action for all stakeholders to reimagine agriculture as a catalyst for sustainable growth and global competitiveness. Together, we can unlock the immense potential of Nigerian agriculture and create a future that benefits all.

CHAPTER 1

Nigeria's Agricultural Landscape

Agriculture has always been a cornerstone of Nigeria's economy, serving as a vital source of livelihood for millions of its citizens. From the lush rainforests of the south to the arid plains of the north, the sector reflects the country's geographical diversity and cultural heritage. Historically, agriculture dominated Nigeria's economic activities, contributing over 60% of the GDP before the discovery of crude oil in the 1950s. During this period, cash crops such as cocoa, groundnuts, palm oil, and cotton were the primary exports, earning Nigeria a prominent position in global agricultural markets. The colonial administration played a significant role in shaping these patterns, establishing large-scale production systems and exporting raw materials to feed industries abroad.

The post-independence era witnessed a continuation of agriculture's importance in the economy. However, the oil boom of the 1970s marked a turning point. The newfound wealth from oil exports shifted attention away from agriculture, leading to a steady decline in investment and productivity. Successive governments prioritized industrialization and urban development, often neglecting the rural farming communities that had sustained the nation for decades. Despite these setbacks, agriculture remains a critical sector, employing approximately 70% of Nigeria's labor force and contributing significantly to food security and rural livelihoods.

Nigeria's agricultural landscape is incredibly diverse, shaped by its unique climatic zones and agro-ecological conditions. The country produces a wide variety of crops, ranging from staples such as cassava, yams, maize, and rice to cash crops like cocoa, rubber, and palm oil. Cassava, in particular, stands out as a staple food for millions and a raw material for industries producing starch, ethanol, and animal feed. Cocoa, predominantly grown in the southwestern states, remains a major export commodity, linking Nigerian farmers to the global chocolate industry. In the north, cereals such as millet, sorghum, and wheat are cultivated extensively, reflecting the region's adaptation to semi-arid conditions. Alongside crop production, Nigeria boasts a vibrant livestock sector, including cattle, goats, sheep, and poultry. Cattle rearing, concentrated in the northern regions, supports both local consumption and trade across West Africa.

The contribution of agriculture to Nigeria's Gross Domestic Product has fluctuated over time. While it no longer dominates the economy as it did in the pre-oil era, agriculture still accounts for about 23-30% of GDP annually. Beyond its economic significance, the sector plays a critical role in providing employment, particularly in rural areas where opportunities are limited. Women and youth are key participants in agricultural activities, from planting and harvesting to processing and marketing. Their involvement underscores the social dimension of agriculture, which serves as a safety net for vulnerable populations and a driver of community development.

Regional variations add complexity to Nigeria's agricultural landscape. In the northern states, the arid and semi-arid climate favors the cultivation of drought-resistant crops like millet and sorghum, as well as extensive livestock farming. The middle belt, often referred to as Nigeria's "food basket," benefits from fertile soils and moderate rainfall, supporting a mix of crops including yams, cassava, maize, and legumes. In contrast, the southern states, with their humid and tropical climate, are ideal for cash crops such as cocoa, oil palm, and rubber. The coastal areas also support fishing, an essential source of protein for many communities. These regional differences highlight the need for tailored agricultural policies and interventions that address the unique challenges and opportunities of each zone.

Despite its potential, Nigeria's agricultural sector faces significant challenges. Infrastructure deficits, such as poor road networks and inadequate storage facilities, hinder the efficient movement of goods from farms to markets. Post-harvest losses remain a major concern, with estimates suggesting that up to 40% of produce is lost due to spoilage and inefficiencies in handling. Climate change adds another layer of complexity, as erratic rainfall patterns and rising temperatures threaten crop yields and livestock productivity. Land tenure issues, limited access to finance, and outdated farming practices further constrain the sector's growth.

Efforts to address these challenges have gained momentum in recent years. Government initiatives like the Agricultural Transformation Agenda (ATA) and the Green Alternative Framework have sought to revitalize the sector by improving productivity, enhancing market access, and promoting private sector participation. These programs emphasize the adoption of modern farming techniques, the use of improved seeds and fertilizers, and the development of value chains that connect farmers to domestic and international markets. While progress has been made, much work remains to be done to unlock the full potential of Nigeria's agricultural sector.

The transformation of Nigeria's agricultural sector requires a multifaceted approach that not only addresses existing challenges but also leverages modern technology, policy reforms, and community-driven initiatives. At the heart of this transformation lies the need for a more efficient and inclusive supply chain system. An optimized supply chain ensures that the journey of agricultural produce from the farm to the consumer's table is seamless, cost-effective, and sustainable. Achieving this requires investments in critical infrastructure, capacity building, and fostering strong partnerships between stakeholders.

Infrastructure remains one of the most pressing bottlenecks in Nigeria's agricultural sector. Poor road networks, particularly in rural areas where the majority of agricultural activities take place, hinder farmers' access to markets. Many roads become impassable during the rainy season, leading to delays, increased transportation costs, and significant post-harvest losses. Beyond roads, the lack of adequate storage facilities exacerbates the issue. Farmers often have to sell their produce immediately after harvest, regardless of market conditions, to avoid spoilage. This practice not only reduces their earnings but also leads to seasonal gluts and price instability. Investing in modern storage technologies, such as cold chains and silos, is crucial to reducing waste and ensuring that produce can reach markets in optimal condition.

The role of technology in modernizing Nigeria's agricultural sector cannot be overstated. Digital platforms are emerging as powerful tools to connect farmers with buyers, streamline logistics, and provide real-time market information. Mobile applications now enable farmers to access weather forecasts, crop advisory services, and even credit facilities. Blockchain technology offers opportunities for enhancing transparency in the supply chain, particularly in tracking produce from farm to market. This is especially relevant for export crops, where traceability and adherence to international standards are critical for competitiveness. However, the adoption of such technologies faces barriers, including low digital literacy among rural farmers and limited internet penetration in remote areas. Addressing these gaps through targeted training programs and investments in rural connectivity is vital.

Policy reforms are another critical pillar for agricultural transformation. Over the years, Nigeria has introduced several policies aimed at boosting agricultural productivity and ensuring food security. While some have yielded positive results, others have been marred by inconsistent implementation and lack of stakeholder engagement. A more cohesive and transparent policy framework is needed to address issues such as land tenure insecurity, which discourages investment in large-scale farming. Land ownership in Nigeria is often governed by traditional systems that limit farmers' ability to secure long-term leases or use their land as collateral for loans. Streamlining land policies to provide clarity and security will empower farmers and attract private sector investment.

Financing remains a significant challenge for many farmers, particularly smallholders who form the backbone of Nigeria's agricultural sector. Despite various government programs to provide subsidized loans and grants, access to affordable credit remains limited. Many farmers rely on informal lending systems with high-interest rates, which often trap them in cycles of debt. Expanding microfinance initiatives, promoting cooperatives, and fostering public-private partnerships can provide more inclusive and sustainable financing options. Furthermore, innovative financial instruments such as weather-indexed insurance can protect farmers from climate-related risks, providing them with a safety net to reinvest in their operations.

Community-driven initiatives also have a critical role to play. Empowering farmer cooperatives and associations can enhance bargaining power, improve access to inputs, and facilitate collective marketing efforts. These groups can also serve as vehicles for knowledge transfer, helping farmers adopt modern techniques and best practices. Capacity building programs should target not only farmers but also other actors in the supply chain, including transporters, processors, and marketers. A well-trained workforce is essential for ensuring that each link in the supply chain operates efficiently and contributes to overall productivity.

Environmental sustainability must be integrated into any discussion about the future of Nigeria's agriculture. Unsustainable farming practices, such as slash-and-burn agriculture, have led to soil degradation and deforestation, reducing the sector's long-term viability. Climate-smart agriculture, which focuses on practices that increase productivity while reducing environmental impacts, offers a path forward. Techniques such as conservation tillage, agroforestry, and integrated pest management can help preserve natural resources while maintaining high yields. Additionally, efforts to promote renewable energy sources, such as solar-powered irrigation systems, can reduce the sector's carbon footprint and improve resilience to climate variability.

The human element is perhaps the most crucial factor in Nigeria's agricultural landscape. Farmers are at the heart of the supply chain, and their empowerment is fundamental to any transformation effort. This includes ensuring that they receive fair prices for their produce and have access to essential services such as healthcare and education. Women, who constitute a significant proportion of the agricultural workforce, must be given equal opportunities to participate in and benefit from agricultural initiatives. Recognizing and addressing gender disparities in access to land, finance, and training will not only improve equity but also enhance overall productivity.

As Nigeria looks to the future, the opportunities for agricultural growth are immense. With a growing population and rising urbanization, the demand for food and agricultural products is set to increase dramatically. Meeting this demand requires a coordinated effort to address the inefficiencies in the current system and build a resilient, market-oriented agricultural sector. Investments in research and development will be crucial for developing high-yield, climate-resilient crop varieties and innovative farming techniques. Strengthening partnerships with international organizations and leveraging global best practices can also accelerate progress.

The future of Nigeria's agricultural sector lies in its ability to create a holistic and interconnected ecosystem that balances production, sustainability, and economic viability. This vision requires breaking away from traditional practices that limit productivity and embracing a future-oriented approach that integrates technology, modern farming methods, and market-oriented strategies. To achieve this, the government, private sector, and international partners must collaborate in creating an enabling environment for innovation and investment.

One of the key areas of focus should be the development of robust agricultural value chains. A value chain encompasses all the activities and processes that add value to a product from the farm to the consumer, including production, processing, storage, transportation, and marketing. In Nigeria, value chains are often fragmented, with weak linkages between farmers and other actors such as processors and exporters. This fragmentation leads to inefficiencies, higher costs, and reduced competitiveness. Strengthening these value chains through targeted interventions, such as the establishment of processing hubs and agricultural clusters, can create a seamless flow of goods and services, thereby improving the overall efficiency and profitability of the sector.

The role of agribusiness cannot be overlooked in driving this transformation. Agribusinesses serve as the bridge between smallholder farmers and larger markets, offering opportunities for scaling production and accessing new markets. In Nigeria, smallholder farmers dominate the agricultural landscape, accounting for over 80% of agricultural output. However, their limited access to resources, knowledge, and technology often constrains their productivity. Encouraging partnerships between smallholders and agribusiness firms can help farmers improve their practices, access better input, and benefit from economies of scale. Contract farming arrangements, where agribusiness firms provide inputs and technical support in exchange for guaranteed produce, have proven successful in several countries and could be adapted to the Nigerian context.

Another critical component is the integration of young people into the agricultural workforce. With a rapidly growing youth population, Nigeria faces the dual challenge of unemployment and an aging farming population. Agriculture offers a viable solution to both problems, but the sector must be made more appealing to young people. This involves shifting perceptions of farming as a labor-intensive and low-paying activity to a dynamic and lucrative profession. Programs that provide training in agribusiness, digital agriculture, and modern farming techniques can empower young people to take on leadership roles in the sector. Additionally, access to finance and mentorship programs can support them in establishing their ventures.

Market access is another pivotal factor in the success of Nigeria's agricultural sector. While domestic demand for food is rising, international markets present immense opportunities for Nigerian produce, particularly cash crops like cocoa, sesame, and ginger. However, accessing these markets requires meeting stringent quality and safety standards. Strengthening quality control systems, investing in certification programs, and providing training on international market requirements can help Nigerian farmers and exporters gain a competitive edge. Furthermore, the establishment of commodity exchanges and marketing boards can stabilize prices and provide farmers with better returns for their efforts.

Infrastructure remains a recurring theme, as it underpins nearly every aspect of agricultural development. Without efficient transportation networks, storage facilities, and energy supply, even the most productive farming systems can fail to deliver results. Rural electrification is particularly crucial for enabling value addition through processing. Farmers who can process their produce into semi-finished or finished goods, such as turning cassava into flour or tomatoes into paste, can significantly increase their earnings. This not only reduces post-harvest losses but also creates employment opportunities along the value chain.

Water resource management is another area that demands urgent attention. Nigeria is blessed with abundant water resources, yet many farmers depend on rain-fed agriculture, leaving them vulnerable to climate variability. Expanding irrigation infrastructure and promoting water conservation techniques can mitigate the risks associated with unpredictable rainfall patterns. Solar-powered irrigation systems, in particular, offer a sustainable solution for regions with limited access to electricity. These systems can enable year-round farming, increasing productivity and reducing the seasonal fluctuations in food supply.

Collaboration and knowledge sharing between farmers, researchers, and policymakers are essential for driving innovation in the agricultural sector. Research institutions must work closely with farmers to develop and disseminate practical solutions to the challenges they face. This includes breeding crop varieties that are high-yielding, disease-resistant, and adapted to local conditions. Extension services play a critical role in this process by acting as a conduit for transferring knowledge and technology from researchers to farmers. However, the current extension system in Nigeria is underfunded and understaffed, leaving many farmers without access to the support they need. Revitalizing this system through increased funding, training, and the use of digital platforms can enhance its reach and effectiveness.

Environmental sustainability must remain at the forefront of agricultural development efforts. As Nigeria seeks to boost productivity, it must also ensure that this growth does not come at the expense of its natural resources. Unsustainable practices, such as overgrazing, deforestation, and excessive use of chemical inputs, have already taken a toll on the environment. Adopting sustainable practices such as agroforestry, crop rotation, and organic farming can help preserve soil fertility, reduce greenhouse gas emissions, and maintain biodiversity. Policymakers must create incentives for farmers to adopt these practices, such as subsidies for organic inputs and payments for ecosystem services.

In the face of these challenges and opportunities, one cannot underestimate the importance of leadership and governance. Strong institutions are needed to coordinate efforts across various stakeholders, from farmers and agribusinesses to development partners and policymakers. Transparent and inclusive governance structures can ensure that resources are allocated effectively and that the benefits of agricultural growth are shared equitably. Corruption and inefficiency must be addressed to build trust and confidence among stakeholders, particularly private investors.

Ultimately, the future of Nigeria's agricultural sector depends on the collective effort of all stakeholders. Farmers, as the primary actors in the supply chain, must be empowered with the tools, knowledge, and resources they need to succeed. Governments at all levels must create policies and infrastructure that support growth and sustainability. The private sector, including agribusinesses and financial institutions, must step up to invest in the sector and drive innovation. Civil society organizations and international partners have a role to play in advocating in inclusive development and providing technical support. By working together, Nigeria can transform its agricultural sector into a model of efficiency, resilience, and sustainability, ensuring food security and economic prosperity for generations to come.

This vision is not merely aspirational, it is achievable with the right mix of determination, investment, and collaboration. The journey from farm to market can be made smoother and more equitable, unlocking the vast potential of Nigeria's agricultural landscape. The challenges are great, but the rewards are even greater, promising a future where agriculture serves as a cornerstone of national development and a source of pride for all Nigerians.

1.1 The Role of Public-Private Partnerships in Agricultural Transformation

Public-private partnerships (PPPs) represent a powerful mechanism for driving growth and innovation in Nigeria's agricultural sector. The challenges facing agriculture ranging from infrastructure deficits to financing gaps are too extensive for any single entity to address effectively. By leveraging the strengths of both the public and private sectors, PPPs can create sustainable solutions that benefit all stakeholders. The public sector, with its regulatory authority and ability to mobilize resources, can provide an enabling environment necessary for private investment. Conversely, the private sector brings expertise, efficiency, and innovation to the table, enabling projects to be implemented more effectively.

In Nigeria, PPPs have already shown promise in addressing critical areas such as storage, processing, and market access. For example, partnerships between government agencies and private agribusiness firms have resulted in the establishment of processing facilities that add value to raw agricultural products. These facilities not only reduce post-harvest losses but also create employment opportunities and enhance the competitiveness of Nigerian produce in both domestic and international markets. Similarly, PPPs have been instrumental in developing transportation networks and logistics hubs that improve the efficiency of supply chains.

One area where PPPs hold significant potential is agricultural financing. By collaborating with financial institutions, the government can design and implement innovative financing models that cater to the unique needs of farmers. For instance, risk-sharing facilities can incentivize banks to lend to smallholder farmers by mitigating the risks associated with agricultural loans. Additionally, partnerships with fintech companies can expand access to digital financial services, enabling farmers to receive payments, access credit, and purchase inputs through their mobile phones.

However, for PPPs to succeed, transparency and accountability must be prioritized. Clear agreements outlining the roles and responsibilities of each party, along with mechanisms for monitoring and evaluation, are essential. The government must also create a stable policy environment that fosters investor confidence. When implemented effectively, PPPs can serve as a catalyst for agricultural transformation, unlocking the sector's potential and driving inclusive economic growth.

1.2 Harnessing Regional Specialization for Competitive Advantage

Nigeria's diverse agro-ecological zones present a unique opportunity to harness regional specialization for greater efficiency and competitiveness. Each region in the country has distinct climatic and soil conditions that favor the production of specific crops and livestock. By focusing on these natural strengths, Nigeria can optimize resource use, reduce production costs, and enhance the quality of its agricultural outputs. Regional specialization also enables the development of targeted value chains, which can boost productivity and create economic opportunities in rural areas.

The northern region, for instance, is well-suited for the cultivation of drought-resistant crops such as millet, sorghum, and ground nuts, as well as extensive livestock farming. In contrast, the southern region's humid tropical climate supports the production of cash crops like cocoa, oil palm, and rubber. The middle belt, often referred to as Nigeria's agricultural heartland, is ideal for staples such as yams, cassava, maize, and rice. By strategically investing in these zones, Nigeria can achieve economies of scale, making its agricultural products more competitive both locally and internationally.

Regional specialization also offers opportunities for developing agro-industrial clusters that serve as hubs for processing, storage, and distribution. These clusters can attract private investment, foster innovation, and facilitate the transfer of knowledge and technology. For example, cocoa-processing hubs in the southwest could produce value-added products such as chocolate and cocoa butter, significantly increasing export revenues. Similarly, livestock hubs in the north could focus on meat processing, leather production, and dairy products, creating a range of value-added goods.

To fully realize the benefits of regional specialization, coordinated efforts are required from both the government and private sector. This includes investing in infrastructure such as roads, storage facilities, and irrigation systems, as well as providing farmers with access to quality inputs, training, and financing. Policies that promote inter-regional trade and collaboration can further strengthen the agricultural sector, ensuring that surplus production in one region meets demand in another.

Ultimately, regional specialization is not just about maximizing output, it's about creating an integrated and sustainable agricultural system that leverages Nigeria's unique strengths. By focusing on what each region does best, Nigeria can build a resilient agricultural sector that contributes to food security, economic growth, and global competitiveness.

CHAPTER 2

Understanding Agricultural Supply Chains

Agricultural supply chains are the lifeblood of the global food system, encompassing the intricate network of activities, people, resources, and processes that bring food from the farm to the consumer's plate. In Nigeria, the agricultural supply chain plays a pivotal role in connecting millions of smallholder farmers to local, regional, and international markets. However, inefficiencies within this system often result in high costs, significant losses, and limited value addition, thereby undermining the sector's potential to drive economic growth and improve livelihoods. This chapter delves into the components of agricultural supply chains, the challenges specific to Nigeria, and how global best practices can provide a blueprint for improvement.

At its core, an agricultural supply chain begins with production on the farm, followed by processing, storage, transportation, distribution, and ultimately, consumption. Each stage involves multiple actors, from farmers and processors to traders, retailers, and consumers. For

Nigeria, where smallholder farmers dominate the agricultural landscape, the supply chain is often fragmented, with weak linkages between these actors. Farmers typically lack direct access to processors and markets, relying instead on intermediaries who often exploit them by paying below-market prices. This disconnect limits farmers' earnings and discourages investment in improved inputs and technology.

One of the most pressing issues in Nigeria's supply chain is the high rate of post-harvest losses. Estimates suggest that up to 40% of perishable produce is lost before it reaches consumers, primarily due to inadequate storage and inefficient transportation. For instance, tomatoes produced in the northern region often spoil before they can be transported to southern markets, resulting in both economic losses for farmers and food shortages for consumers. Addressing this issue requires investment in cold storage facilities, improved packaging techniques, and reliable transportation networks. Additionally, training farmers and other supply chain actors in best practices for handling and preserving produce can further reduce losses.

Processing and value addition represent another weak link in Nigeria's supply chain. The country exports many of its agricultural products in raw form, foregoing the higher revenues associated with processed goods. For example, while Nigeria is one of the largest producers of cocoa, much of it is exported as raw beans rather than processed into chocolate or cocoa butter. This reliance on raw exports not only reduces profitability but also limits job creation and industrial development. By developing local processing industries, Nigeria can capture more value within the supply chain and build a more resilient agricultural economy.

The transportation and logistics segment of the supply chain is equally critical. In many parts of Nigeria, poor road infrastructure and unreliable transportation systems hinder the efficient movement of goods from farms to markets. Farmers in remote areas often face significant challenges in accessing urban markets, where demand and prices are higher. This disconnect not only restricts farmers' income but also leads to regional imbalances in food availability. Developing a robust logistics network, including roads, railways, and waterways, can improve market connectivity and enhance supply chain efficiency. Furthermore, introducing digital logistics platforms can help match supply with demand in real time, reducing waste and optimizing resource use.

A key enabler of efficient agricultural supply chains is technology. From digital platforms that connect farmers directly to buyers to mobile applications that provide market information, technology has the potential to revolutionize Nigeria's agricultural supply chains. For example, e-commerce platforms like Farm rowdy have successfully linked farmers with end-users, eliminating intermediaries and ensuring fair prices. Similarly, innovations in blockchain technology can enhance traceability, ensuring that products meet quality and safety standards required for export. However, widespread adoption of these technologies requires addressing barriers such as low digital literacy and limited access to the internet in rural areas.

Financing is another critical aspect of the supply chain. Many smallholder farmers and supply chain actors lack access to affordable credit, which constrains their ability to invest in inputs, equipment, and infrastructure. Traditional financial institutions are often reluctant to lend to farmers due to perceived risks and the lack of collateral. Innovative financing mechanisms, such as value chain financing and agricultural credit guarantees, can bridge this gap. For instance, processors and buyers can provide pre-harvest loans to farmers, to be repaid in produce, creating a mutually beneficial arrangement.

Collaboration and coordination among stakeholders are essential for building resilient supply chains. Farmers, processors, traders, and policymakers must work together to identify bottlenecks and develop solutions that benefit the entire system. Cooperatives and farmer associations can play a significant role in this regard, providing farmers with collective bargaining power, access to resources, and opportunities for knowledge sharing. Similarly, partnerships between the public and private sectors can address systemic challenges, such as infrastructure development and regulatory reform.

Global best practices offer valuable insights into improving agricultural supply chains in Nigeria. Countries like India and Brazil, which have faced similar challenges, have implemented successful strategies to enhance efficiency and inclusivity. For example, India's National Agricultural Market (eNAM) platform has transformed the country's agricultural supply chains by connecting farmers to a unified digital marketplace, reducing transaction costs and ensuring better prices.

Nigeria can draw lessons from such initiatives, adapting them to its unique context.

To fully optimize agricultural supply chains in Nigeria, there is a pressing need to deepen investments and focus on innovation across multiple interconnected domains. Strengthening supply chains requires moving beyond surface-level fixes and addressing systemic inefficiencies at each stage. This means creating an ecosystem where farmers, processors, logistics providers, and markets work in synergy, supported by technology, infrastructure, and regulatory frameworks that prioritize growth and inclusivity.

A critical area for expansion is the development of agro-industrial zones and clusters. These zones provide centralized hubs where agricultural produce can be aggregated, processed, and packaged before distribution. By situating these hubs near production areas, Nigeria can significantly reduce transportation costs and time, while also ensuring that perishable goods are processed quickly to minimize losses. Agro-industrial clusters also attract investment from private firms by creating a business-friendly environment with access to shared infrastructure such as power, water, and cold storage. These clusters can serve as catalysts for rural industrialization, creating jobs and enhancing the value chain. Examples from countries like Ethiopia, where the government established agro-industrial parks to boost exports, illustrate the potential impact of such initiatives in Nigeria.

Another focus should be the integration of digital technologies that promote transparency, traceability, and efficiency across the supply chain. Blockchain technology, for example, can revolutionize the way agricultural products are tracked from farm to market. By creating a digital ledger that records every transaction and movement of produce, blockchain ensures transparency and builds trust among stakeholders. For export-oriented crops like cocoa and sesame, traceability is critical to meeting international quality standards and gaining access to premium markets. Digital platforms can also improve coordination within the supply chain, enabling real-time matching supply with demand and optimizing resource allocation.

Policy reforms remain a cornerstone of supply chain transformation. The Nigerian government must address regulatory bottlenecks that impede the free flow of goods. For instance, excessive taxation and multiple checkpoints along transport routes increase costs and discourage investment. Simplifying these processes, while enforcing quality and safety standards, can enhance the competitiveness of Nigerian agricultural products. Additionally, creating a harmonized national framework for agriculture-related taxes, incentives, and trade policies will encourage private sector participation and reduce inefficiencies.

Equally important is the role of capacity building and knowledge dissemination. Farmers and supply chain actors must be equipped with the skills and knowledge needed to adapt to modern practices and technologies. Extension services, which act as a bridge between research institutions and farmers, need to be revitalized and expanded.

Digital tools can complement traditional extension services by providing farmers with access to training materials, market information, and advisory services through their mobile phones. Moreover, partnerships with academic institutions and non-governmental organizations can enhance the research and development of innovative solutions tailored to Nigeria's agricultural challenges.

Collaboration among stakeholders is the backbone of a resilient supply chain. Farmer cooperatives and producer associations can play a pivotal role in aggregating produce, negotiating better prices, and providing members with access to shared resources such as equipment and storage facilities. These groups also provide a platform for collective advocacy, enabling farmers to influence policies that affect their livelihoods. On a larger scale, partnerships between the public and private sectors can drive systemic changes, such as developing infrastructure, promoting financial inclusion, and fostering innovation. International development agencies and donor organizations can further support these efforts by providing technical assistance and funding for pilot projects and capacity-building programs.

Lastly, sustainability must be a guiding principle in the design and implementation of agricultural supply chains. Climate change poses a significant threat to agricultural productivity in Nigeria, with increasing incidences of droughts, floods, and erratic weather patterns. Integrating climate-smart agriculture into the supply chain can help mitigate these risks and ensure long-term viability. Practices such as conservation agriculture, agroforestry, and the use of renewable energy in processing and storage can reduce the environmental footprint of agriculture while maintaining high productivity levels. Additionally, policies that incentivize sustainable practices, such as subsidies for organic farming inputs and carbon credits for reducing greenhouse gas emissions, can encourage adoption among farmers and agribusinesses.

The transformation of Nigeria's agricultural supply chains is not an isolated endeavor—it is part of a broader effort to position agriculture as a cornerstone of national development. A well-functioning supply chain not only benefits farmers and agribusinesses but also contributes to food security, reduces poverty, and drives industrial growth. By addressing inefficiencies, fostering innovation, and promoting sustainability, Nigeria can build a supply chain system that meets the demands of a growing population and positions the country as a major player in the global agricultural market. The journey to achieving this vision is challenging, but the rewards—economic growth, job creation, and improved livelihoods—make it an endeavor worth pursuing. This sets the stage for the next chapter, which will delve into the critical role of infrastructure and logistics in transforming Nigeria's agricultural supply chain landscape.

To fully harness the potential of agricultural supply chains in Nigeria, a strategic focus on the synergy between infrastructure, logistics, and value creation is essential. As the backbone of any efficient supply chain, infrastructure determines the speed, cost, and reliability with which agricultural goods move from production to consumption. Nigeria's current infrastructure deficit not only hinders economic growth but also amplifies the inefficiencies within the agricultural sector, creating significant challenges for farmers, processors, and marketers alike.

The development of robust transportation networks should be a top priority. Poor road conditions in rural areas make it difficult for farmers to transport their produce to markets, often resulting in high costs and extended delays. For instance, during peak harvest seasons, crops like tomatoes, yams, and cassava may perish while waiting for transportation, leading to substantial post-harvest losses. Addressing this issue requires investment in constructing and maintaining rural roads, bridges, and access routes that connect farming communities to urban centers. Beyond road networks, Nigeria's underutilized rail and inland waterway systems present untapped opportunities for efficient bulk transportation of agricultural goods. Revitalizing these systems, alongside integrating multimodal logistics solutions, can significantly reduce transportation costs and enhance market access.

Storage infrastructure also plays a critical role in ensuring the quality and availability of agricultural products throughout the year. Inadequate storage facilities, especially for perishable goods, exacerbate the problem of seasonal gluts and price volatility. Establishing modern storage solutions, such as cold storage units, silos, and climate-controlled warehouses, can mitigate these issues. Government incentives and public-private partnerships can encourage investment in storage infrastructure, especially in high-yield agricultural regions. Moreover, promoting small-scale, on-farm storage technologies, like hermetic bags and solar dryers, can empower farmers to preserve their produce and negotiate better prices during off-peak seasons.

Logistics innovation is equally critical for overcoming inefficiencies in the supply chain. Traditional systems of aggregating and distributing agricultural goods are often disorganized and prone to delays. Digital logistics platforms, which connect farmers with buyers and transporters, can streamline these processes and reduce transaction costs. For example, mobile apps that allow farmers to book transportation or find nearby markets in real-time can optimize resource allocation and prevent wastage. Additionally, the integration of data analytics and artificial intelligence into logistics planning can help predict demand patterns, allocate resources efficiently, and minimize transportation bottlenecks.

Another key element is the establishment of agro-processing hubs in strategic locations across Nigeria. These hubs can serve as value addition centers where raw produce is processed into finished or semi-finished goods, ready for consumption or export. For instance, cassava can be processed into starch, flour, or ethanol, while cocoa can be transformed into chocolate and cocoa butter. By reducing the reliance on exporting raw agricultural products, Nigeria can capture more value within its borders, generate foreign exchange, and create employment opportunities. These hubs also provide a platform for training and technology transfer, enabling farmers and processors to adopt the best practices and improve their productivity.

Policy alignment and regulatory reforms must complement infrastructure and logistics development. A cohesive national policy framework that integrates agricultural development, infrastructure planning, and trade facilitation is essential for achieving a seamless supply chain. Policies that reduce bureaucratic bottlenecks, such as unnecessary checkpoints on transport routes and inconsistent taxation across states, can significantly enhance efficiency. Similarly, creating export-friendly policies, such as simplified customs procedures and incentives for value-added exports, can open up new markets for Nigerian agricultural products.

Technology and innovation will play a transformative role in achieving these goals. Digital marketplaces can eliminate intermediaries, allowing farmers to connect directly with buyers and receive fairer prices. Blockchain technology can ensure transparency and traceability, which are critical for meeting international standards and building trust among stakeholders. Furthermore, geospatial technologies, such as satellite imaging and GPS mapping, can support better planning and monitoring of infrastructure projects, ensuring they align with agricultural priorities.

Capacity building remains a cornerstone of sustainable supply chain development. Farmers, transporters, and processors need continuous training to adapt to new technologies, infrastructure, and market demands. Collaborative efforts between government agencies, academic institutions, and development organizations can create training programs that enhance skills and knowledge across the supply chain. For example, training programs in post-harvest handling, quality assurance, and logistics management can significantly reduce losses and improve efficiency.

Environmental sustainability must also be integrated into supply chain development. Infrastructure projects should be designed to minimize their ecological impact while promoting climate resilience. For example, using renewable energy sources such as solar power for cold storage units and processing facilities can reduce greenhouse gas emissions and lower operational costs. Similarly, adopting sustainable construction practices, such as using eco-friendly materials for road and bridge projects, can contribute to long-term environmental preservation.

2.1 Enhancing Agricultural Supply Chains Through Technology

The integration of technology into agricultural supply chains has the potential to revolutionize Nigeria's agricultural sector, addressing inefficiencies and creating new opportunities for growth. From digital platforms to precision farming techniques, technology offers innovative solutions that can streamline processes, reduce costs, and improve the overall quality of agricultural products. For a country like Nigeria, where many supply chains remain fragmented and informal, the adoption of technology is not just an option but a necessity for achieving a competitive and sustainable agricultural economy.

Digital platforms are one of the most transformative tools for connecting various actors in the supply chain. These platforms enable farmers to access real-time market information, connect with buyers, and arrange logistics without relying on intermediaries. For instance, mobile applications like Farmcrowdy and Thrive Agric have empowered Nigerian farmers to raise capital, access markets, and improve their practices. Similarly, e-commerce platforms provide opportunities for farmers to sell directly to consumers, by passing traditional market systems that often exploit them. Such platforms ensure fairer pricing, reduce post-harvest losses, and enhance farmers' bargaining power.

Blockchain technology is another promising innovation that can enhance transparency and traceability in supply chains. By creating a decentralized ledger of transactions, blockchain ensures that every step in the supply chain, from farm to market, is documented and verifiable. This is particularly critical for export commodities such as cocoa, sesame seeds, and cashew nuts, where compliance with international standards is essential. Blockchain not only builds trust among stakeholders but also opens access to premium markets that demand traceability and quality assurance.

Precision farming, enabled by technologies like drones, sensors, and satellite imaging, can improve the efficiency of production processes. Farmers can use these tools to monitor crop health, optimize water usage, and apply fertilizers more accurately, reducing waste and increasing yields. Combined with data analytics, precision farming allows for better decision-making and resource allocation, ensuring that farmers maximize productivity while minimizing environmental impact.

Despite these opportunities, barriers to technology adoption remain significant. Limited internet penetration in rural areas, low digital literacy among farmers, and the high cost of technology solutions are key challenges. Addressing these requires investments in digital infrastructure, targeted training programs, and public-private partnerships to subsidize technology access for smallholder farmers. With the right support, technology can become the backbone of a modernized agricultural supply chain system in Nigeria.

2.2 Building Sustainable Supply Chains for Long-Term Growth

Sustainability is an essential consideration in the transformation of Nigeria's agricultural supply chains. As the sector grows to meet increasing demand, it is crucial to ensure that this growth does not come at the expense of the environment, or the well-being of the communities involved. Sustainable supply chains prioritize practices that protect natural resources, reduce waste, and promote social equity, creating a balanced approach to economic development and environmental stewardship.

One of the most pressing environmental challenges in Nigeria's agricultural supply chains is the overuse of natural resources, including land and water. Unsustainable farming practices, such as deforestation and excessive use of chemical fertilizers, have led to soil degradation and reduced agricultural productivity. Integrating climate-smart agriculture into supply chains can help address these issues. Practices such as agroforestry, crop rotation, and conservation tillage preserve soil fertility and reduce greenhouse gas emissions while maintaining high productivity levels. Furthermore, the adoption of renewable energy sources, such as solar-powered irrigation and processing facilities, can lower the carbon footprint of supply chains and enhance resilience to climate change.

Waste management is another critical aspect of sustainability. Post-harvest losses, which account for up to 40% of Nigeria's agricultural output, represent not only an economic loss but also an environmental burden. Reducing these losses through improved storage, transportation, and processing methods can significantly enhance supply chain efficiency and sustainability. Innovations like biodegradable packaging and recycling programs can further minimize waste and promote a circular economy within the agricultural sector.

Social equity is equally important in building sustainable supply chains. Empowering smallholder farmers, women, and youth ensures that the benefits of supply chain improvements are shared equitably. This includes providing access to training, financing, and markets, as well as creating inclusive policies that address gender and social disparities. Fair trade practices and certifications can also enhance the social sustainability of supply chains, ensuring that workers receive fair wages and operate under safe conditions.

In conclusion, sustainability is not just an add-on but a core principle that must guide the development of Nigeria's agricultural supply chains. By prioritizing environmental conservation, waste reduction, and social inclusion, Nigeria can build a supply chain system that supports long-term growth while preserving its natural and social capital. This sets the stage for exploring how investments in sustainability can drive innovation and competitiveness in the global agricultural market.

CHAPTER 3

Infrastructure and Logistics in Agricultural Supply Chains

Infrastructure and logistics are the twin pillars of a functional and efficient agricultural supply chain. In Nigeria, the lack of adequate infrastructure and fragmented logistics systems significantly undermines the agricultural sector's ability to reach its full potential. Poor road networks, insufficient storage facilities, unreliable transportation, and limited access to energy collectively contribute to high costs, post-harvest losses, and inefficiencies. Addressing these challenges is essential for creating a seamless connection between farmers, markets, and consumers.

Agricultural supply chains rely heavily on transportation infrastructure to move goods from rural production areas to urban consumption centers. However, Nigeria's road networks, particularly in rural areas, are in a state of disrepair. Many farmers face difficulties in transporting their produce to markets, especially during the rainy season when roads often become impassable. This lack of access not only increases transportation costs but also limits farmers' ability to sell their produce at competitive prices. Investments in constructing and maintaining rural roads, coupled with the development of feeder roads that connect remote farming communities to major highways, can greatly improve market access and reduce costs.

Beyond roads, Nigeria's underutilized rail and inland waterway systems present significant opportunities for efficient bulk transportation of agricultural goods. Rail transport, in particular, offers a cost-effective solution for moving large volumes of produce over long distances. Revitalizing the rail network and integrating it with road and waterway systems can create a multimodal logistics framework that enhances the efficiency of agricultural supply chains. For instance, transporting grains from northern Nigeria to southern markets via rail can reduce transit times and minimize spoilage, benefiting both farmers and consumers.

Storage infrastructure is another critical component of agricultural logistics. Without adequate storage facilities, farmers are often forced to sell their produce immediately after harvest, regardless of market conditions. This results in seasonal gluts and low prices for farmers, while consumers face high prices during off-peak periods. Developing modern storage solutions, such as cold storage units for perishable goods and silos for grains, can help stabilize prices and ensure year-round availability of agricultural products. Small-scale, on-farm storage technologies, such as hermetic bags and solar dryers, also empower farmers to preserve their produce and reduce post-harvest losses.

Energy access is a key enabler for modernizing infrastructure and logistics. Many rural farming communities lack reliable electricity, which is essential for operating cold storage facilities, processing equipment, and logistics systems. Expanding rural electrification through renewable energy sources, such as solar and wind power, can provide a sustainable and cost-effective solution. For example, solar-powered cold storage units can extend the shelf life of perishable goods, allowing farmers to access more distant markets and secure better prices for their produce.

In addition to physical infrastructure, digital logistics platforms can enhance the efficiency of agricultural supply chains. These platforms connect farmers with transporters, buyers, and processors, enabling real-time coordination and resource optimization. For instance, mobile apps that allow farmers to book transportation services or find buyers for their product can reduce transaction costs and prevent delays. Data analytics and artificial intelligence can further improve logistics planning by predicting demand patterns, optimizing routes, and minimizing waste.

The private sector has a critical role to play in addressing infrastructure and logistics challenges. Public-private partnerships (PPPs) can mobilize resources and expertise for large-scale infrastructure projects, such as the construction of processing hubs and logistics centers. These partnerships can also promote innovation by encouraging private companies to invest in technologies that enhance supply chain efficiency. For example, agritech startups focusing on digital logistics solutions have the potential to disrupt traditional systems and create new opportunities for farmers and businesses.

Policy reforms are essential for creating an enabling environment for infrastructure and logistics development. Streamlining regulations, reducing bureaucratic bottlenecks, and eliminating multiple taxation points along transport routes can lower costs and encourage investment. Additionally, integrating infrastructure planning with agricultural policies can ensure that investments are targeted toward areas with the highest impact on productivity and market access.

To truly optimize Nigeria's agricultural supply chains, addressing infrastructure and logistics challenges requires a holistic approach that integrates strategic planning, innovation, and stakeholder collaboration. While the preceding discussion highlights key areas such as transportation, storage, and energy, a deeper exploration of systemic solutions and future opportunities is essential for charting a sustainable path forward.

One of the most impactful strategies for transforming agricultural supply chains in Nigeria is the development of integrated infrastructure networks that connect production zones to markets efficiently and cost-effectively. An integrated network involves not only roads, railways, and waterways but also strategically located hubs for aggregation, processing, and distribution. These networks reduce the fragmentation that currently characterizes supply chains, where goods often pass through multiple intermediaries before reaching their final destination.

A promising model is the creation of agro-industrial corridors dedicated infrastructure routes that connect high-production areas with processing facilities and major market centers. Such corridors can prioritize investments in roads, rail links, and storage facilities, ensuring that agricultural goods move seamlessly along the supply chain. For example, a corridor linking the grain-producing northern states to Lagos and other coastal cities can facilitate the efficient movement of bulk commodities for domestic consumption and export. Furthermore, incorporating logistics hubs within these corridors can streamline operations by serving as centralized points for grading, sorting, packaging, and storage.

Efforts to modernize Nigeria's rail and waterway systems can also unlock significant benefits. Rail transport offers a cost-effective solution for bulk transportation, particularly for non-perishable items like grains, beans, and tubers. Waterways, such as the Niger and Benue rivers, can be leveraged to transport agricultural goods to coastal markets, reducing road congestion and lowering costs. The development of river ports and storage facilities along these waterways would further enhance their utility, making them integral components of the supply chain.

Logistics optimization goes beyond physical infrastructure; it also involves improving the coordination and management of resources across the supply chain. In Nigeria, where logistics systems are often inefficient and prone to delays, innovative approaches can transform the way agricultural goods are transported, stored, and distributed.

Digital logistics platforms are increasingly gaining traction as a tool for optimizing supply chains. These platforms enable real-time communication between farmers, transporters, processors, and buyers, reducing inefficiencies and ensuring timely delivery of goods. For instance, mobile applications that allow farmers to book transportation services or find nearby markets can significantly reduce waiting times and transportation costs. Platforms like Twiga Foods in Kenya, which aggregates produce from smallholder farmers and delivers it directly to retailers, offer a model that could be adapted to Nigeria's context.

The use of advanced technologies such as GPS tracking and route optimization algorithms can further enhance logistics efficiency. GPS tracking allows transporters and logistics managers to monitor vehicle movements and ensure timely deliveries, while route optimization algorithms identify the most efficient paths for transporting goods, reducing fuel costs and travel time. Combining these technologies with predictive analytics can enable supply chain actors to anticipate demand patterns, plan inventory levels, and allocate resources effectively.

Automation is another frontier for logistics innovation. Automated warehousing systems, equipped with robotics and conveyor belts, can speed up the sorting and packaging of agricultural goods, particularly in high-volume processing hubs. Drones are also being explored for last-mile delivery in remote or inaccessible areas, offering a fast and cost-effective solution for reaching underserved communities.

Public-private partnerships (PPPs) can accelerate the adoption of these innovations by bringing together government resources and private sector expertise. For instance, the government can provide subsidies or tax incentives to companies that invest in logistics technologies, while private firms can develop and operate digital platforms or automated systems. Collaboration with international organizations and development partners can also facilitate the transfer of knowledge and best practices from other countries.

To fully realize the potential of improved infrastructure and logistics in Nigeria's agricultural supply chains, long-term sustainability and inclusivity must be prioritized alongside efficiency and innovation. Addressing these challenges on a scale requires not only financial investment but also a comprehensive framework that fosters collaboration among stakeholders, promotes environmental resilience, and builds local capacity to manage and sustain these systems.

3.1 Fostering Collaboration Across Stakeholders

Collaboration is essential for overcoming the systemic challenges that plague Nigeria's agricultural supply chains. At the core of this effort lies the need for coordinated action between the government, private sector, development agencies, and farmer cooperatives. Each of these stakeholders plays a distinct yet interconnected role in creating a functional and resilient supply chain ecosystem.

The government must take the lead in creating an enabling policy and regulatory environment that supports infrastructure and logistics development. This includes reducing bureaucratic bottlenecks, harmonizing transportation-related policies across states, and incentivizing investments in critical infrastructure. For example, establishing a central body to oversee agricultural supply chain development can ensure that resources are allocated strategically and that projects align with national priorities. Transparent procurement processes and performance monitoring are also critical for building trust and ensuring accountability.

The private sector has a unique ability to drive innovation and efficiency in logistics. Agribusinesses and logistics companies can develop and operate modern facilities such as cold storage units, processing hubs, and distribution centers. Additionally, partnerships with technology companies can enhance supply chain management through digital platforms and data analytics tools. Programs such as Build-Operate-Transfer (BOT) models can encourage private companies to invest in large-scale infrastructure projects while ensuring eventual ownership and control by local governments or communities.

Farmer cooperatives and producer associations also have a vital role to play. By aggregating produce, these groups can reduce logistical costs and increase bargaining power for smallholder farmers. Cooperatives can also manage shared infrastructure, such as warehouses and transportation equipment, making these resources accessible to individual farmers who would otherwise be unable to afford them. Development agencies and non-governmental organizations (NGOs) can support these efforts by providing technical assistance, funding, and training programs that strengthen the capacity of cooperatives to manage logistics effectively.

3.2 Promoting Sustainability and Environmental Resilience

Sustainability must be at the forefront of any effort to transform Nigeria's agricultural infrastructure and logistics. The rapid expansion of transportation networks, storage facilities, and processing hubs must be balanced with environmental considerations to avoid exacerbating climate change and resource degradation. By adopting environmentally friendly practices, Nigeria can ensure that its agricultural growth is both resilient and sustainable.

Transportation infrastructure, for instance, can be designed to minimize its environmental footprint. The construction of rural roads and bridges should incorporate sustainable materials and techniques that reduce erosion and maintain soil integrity. Additionally, integrating renewable energy sources, such as solar-powered streetlights along rural roads, can enhance safety and reduce dependence on fossil fuels. Rail and waterway systems also provide a more environmentally friendly alternative to road transport, as they emit fewer greenhouse gases per ton of goods transported.

Cold storage facilities and processing hubs must prioritize energy efficiency and renewable energy integration. Solar-powered cold chains, for example, are already being used in several developing countries to preserve perishable goods while reducing energy costs.

These facilities can also be designed to recycle waste materials, such as turning organic byproducts from processing plants into compost or bioenergy. Promoting a circular economy within agricultural supply chains not only reduces waste but also creates additional revenue streams for farmers and processors.

Climate-smart logistics practices can further enhance sustainability. For instance, optimizing transportation routes and schedules can reduce fuel consumption and emissions, while implementing no-idling policies for delivery vehicles can minimize air pollution. Investment in electric vehicles for short-distance transportation is another area with significant potential, particularly as battery technologies become more affordable.

Community engagement is crucial for ensuring that infrastructure projects align with local needs and priorities. Farmers and rural communities should be involved in the planning and implementation of logistics projects, ensuring that these initiatives address their unique challenges and leverage their local knowledge. Training programs on sustainable practices, such as efficient water use and eco-friendly packaging, can empower farmers and supply chain actors to contribute to environmental conservation.

In conclusion, fostering collaboration and promoting sustainability are integral to the long-term success of Nigeria's agricultural infrastructure and logistics systems. By working together and adopting environmentally responsible practices, stakeholders can create a supply chain ecosystem that not only meets current needs but also supports future generations. These efforts will strengthen Nigeria's position as a leader in agricultural innovation and sustainability, setting the stage for the next chapter on technology's transformative impact on the sector.

CHAPTER 4

Technology and Innovation in Agricultural Supply Chains

Technology and innovation are reshaping agricultural supply chains worldwide, and Nigeria stands at the threshold of a technological revolution in its agricultural sector. With over 70% of the population engaged in agriculture, integrating modern technology into supply chains offers a transformative opportunity to address inefficiencies, reduce waste, and improve productivity. This chapter explores the role of technology in optimizing agricultural supply chains in Nigeria, highlighting key innovations, challenges, and pathways for scalable adoption.

Technology enhances every stage of the agricultural supply chain, from production and processing to transportation and market access. One of the most impactful innovations in recent years has been the rise of digital platforms connecting farmers with buyers, transporters, and suppliers. These platforms eliminate intermediaries, ensuring fair prices for farmers and reducing transaction costs. Mobile applications such as Farmcrowdy and Thrive Agric have enabled farmers to crowdfund investments, receive market information, and access advisory services. These tools empower farmers to make informed decisions, improving yields and profitability.

Blockchain technology is another game-changer for agricultural supply chains. This decentralized ledger system ensures transparency and traceability by recording every transaction along the supply chain. For export crops like cocoa, sesame, and cashews, blockchain can verify quality, ensure compliance with international standards, and enhance trust among buyers. By improving traceability, blockchain technology also opens access to premium markets where consumers demand ethical sourcing and high-quality products.

Precision agriculture, supported by innovations like drones, sensors, and satellite imaging, allows farmers to monitor soil health, crop conditions, and weather patterns in real-time. This data-driven approach minimizes resource wastage, such as water and fertilizers, while maximizing yields. For instance, drones equipped with imaging technology can identify areas of pest infestation or nutrient deficiency, enabling targeted interventions. These technologies are especially valuable in Nigeria, where many farmers rely on traditional methods and face challenges like soil degradation and erratic weather patterns.

Artificial intelligence (AI) and machine learning are also gaining traction in supply chain management. AI-powered systems can analyze large datasets to predict demand, optimize inventory, and identify bottlenecks in logistics. For example, predictive analytics can help farmers plan planting and harvesting schedules based on market trends, reducing the risk of oversupply and price crashes. AI can also enhance route optimization for transportation, ensuring timely delivery of goods and reducing fuel costs.

Despite these advancements, barriers to technology adoption remain significant in Nigeria. Limited internet penetration, particularly in rural areas, hinders access to digital tools and platforms. Many farmers also lack the technical skills required to operate modern technologies, while the high cost of equipment such as drones and sensors remains prohibitive for smallholder farmers. Addressing these challenges requires targeted investments in digital infrastructure, training programs, and affordable technology solutions tailored to the needs of small-scale farmers.

Public-private partnerships (PPPs) can play a pivotal role in overcoming these barriers. By collaborating with technology companies, agribusinesses, and international development agencies, the government can facilitate the rollout of affordable and scalable technological solutions. For example, subsidies or financing schemes can make drones, sensors, and other equipment accessible to farmers. Meanwhile, public investments in internet connectivity and digital literacy programs can bridge the digital divide, enabling broader adoption of technology.

Another promising area for innovation is the development of e-commerce platforms that enable farmers to sell their produce directly to consumers. These platforms, such as Jumia Foods or local farmer markets, can disrupt traditional supply chains by reducing the role of intermediaries and connecting farmers to urban markets. Integrating logistics services into these platforms ensures seamless delivery, reducing waste and ensuring freshness for perishable goods.

Technology also has the potential to enhance financial inclusion within the agricultural sector. Mobile money platforms and digital wallets enable farmers to receive payments securely and access financial services such as loans and insurance. Innovative financing models, such as value chain financing and crowdfunding, can provide farmers with the capital needed to invest in inputs, technology, and infrastructure. Weather-indexed insurance, facilitated through digital platforms, offers a safety net for farmers facing climate-related risks, reducing their vulnerability and enabling long-term planning.

The environmental benefits of technology in agricultural supply chains are equally significant. Renewable energy solutions, such as solar-powered irrigation and processing facilities, reduce reliance on fossil fuels and lower carbon emissions. Smart irrigation systems, driven by data and sensors, optimize water use, ensuring sustainability in water-scarce regions. Similarly, waste-to-energy technologies can convert agricultural byproducts into bioenergy, creating a circular economy within the supply chain.

To fully harness the potential of technology in Nigeria's agricultural supply chains, a comprehensive approach is needed that integrates innovation, capacity building, and systemic change. Technology, while transformative, must be accompanied by supportive policies, infrastructure investments, and educational initiatives to ensure its equitable and effective adoption. Scaling these solutions requires a deeper focus on specific enablers and the practical pathways through which technology can impact every level of the supply chain.

For technology to truly revolutionize Nigeria's agricultural supply chains, stakeholders across the value chain must be equipped with the skills and knowledge necessary to use these innovations effectively. Farmers, particularly smallholders who form the backbone of the agricultural sector, require training to adopt and implement digital tools, precision farming techniques, and modern logistics systems. Many farmers in Nigeria still rely on traditional practices and may view technology as complex or inaccessible. Addressing these perceptions through well-structured, accessible training programs is critical.

Government agencies, NGOs, and private companies can collaborate to provide digital literacy programs tailored to the agricultural context. These initiatives should include hands-on training in using mobile applications, understanding blockchain systems, and employing data-driven decision-making tools. Extension workers, who already serve as a vital link between farmers and modern practices, can be equipped with technology to disseminate knowledge more effectively. For instance, mobile-enabled extension workers can provide real-time advice on crop health, pest control, and weather conditions, increasing productivity and reducing losses.

Peer-to-peer learning and cooperative networks can further accelerate technology adoption. By enabling early adopters to share their experiences and successes with other farmers, these networks foster trust and encourage widespread implementation of innovative solutions. Incorporating technology training into existing farmer associations and cooperatives ensures that these programs are both scalable and sustainable. Moreover, prioritizing the inclusion of women and youth in these initiatives can address long-standing inequities in the sector while leveraging the energy and creativity of younger generations to drive innovation.

A significant challenge in integrating technology into Nigeria's agricultural supply chains is the high cost of many innovations. Drones, IoT sensors, and blockchain systems, while effective, often remain out of reach for smallholder farmers due to limited financial resources. To overcome this barrier, scaling affordable technology solutions must be a key priority. This can be achieved through the development of low-cost alternatives, collaborative financing models, and public-private partnerships (PPPs).

One promising approach is the development of shared technology models. Cooperatives and agribusiness hubs can invest in expensive equipment, such as drones and sensors, which are then rented out to farmers at affordable rates. This sharing economy model reduces the upfront costs for individual farmers while ensuring that they still benefit from advanced technology. Similarly, community-based solar-powered cold storage units can serve multiple farmers, extending the shelf life of perishable goods and preventing waste.

Government subsidies and tax incentives can also make technology more accessible. By reducing the cost of importing agricultural technology or manufacturing it locally, policymakers can lower the financial barriers for farmers and agribusinesses. Additionally, partnerships with technology companies and development agencies can facilitate the donation or subsidized sale of equipment, particularly during the pilot phases of new initiatives.

Innovative financing models are another avenue for scaling technology adoption. Microfinance institutions, crowdfunding platforms, and value chain financing mechanisms can provide farmers with the capital needed to invest in technology. For example, agritech firms can offer farmers equipment on a lease-to-own basis, where payments are deducted from future earnings. Weather-indexed insurance and digital credit scoring systems can further mitigate risks and enhance farmers' access to loans.

Finally, creating a vibrant innovation ecosystem is essential for fostering the development of context-specific solutions. Supporting local startups and research institutions in designing affordable, scalable agricultural technologies tailored to Nigeria's needs can create a pipeline of innovations that address specific challenges within the supply chain. Hackathons, incubator programs, and innovation hubs focused on agriculture can bring together technologists, entrepreneurs, and farmers to co-create solutions that are both practical and impactful.

To ensure the successful integration of technology into Nigeria's agricultural supply chains, fostering an ecosystem that supports continuous innovation, scalability, and adaptability is critical. This effort requires alignment between technological advancements and the realities of Nigeria's agricultural landscape, coupled with robust collaboration across stakeholders. By scaling these efforts and addressing potential bottlenecks, technology can serve as a powerful catalyst for transforming agricultural supply chains and driving inclusive economic growth.

4.1 Strengthening Ecosystems for Agri-Tech Innovation

The development of a thriving Agri-tech ecosystem is essential to drive the adoption of technology in Nigeria's agricultural supply chains. Such an ecosystem involves a network of innovators, investors, policymakers, academic institutions, and end-users working collaboratively to create and scale solutions tailored to Nigeria's unique agricultural challenges. At the heart of this effort lies the need to foster research and development (R&D) initiatives that generate context-specific technologies and innovations.

Local R&D efforts can address the specific needs of Nigeria's diverse agro-ecological zones, ensuring that technologies are adaptable to varying climates, soil types, and farming practices. For example, developing drought-resistant seed varieties for arid northern regions or creating cost-effective irrigation systems for areas prone to water scarcity can significantly improve productivity. Research institutions and universities can also play a crucial role in testing and validating these technologies, ensuring their effectiveness and scalability.

Startups and small and medium enterprises (SMEs) are often at the forefront of innovation, designing practical and affordable solutions for farmers and supply chain actors. Supporting these enterprises through incubators, accelerators, and access to funding can enhance their capacity to bring innovative products to market. Government programs and private-sector initiatives that provide seed funding, mentorship, and networking opportunities can further stimulate the growth of these ventures.

Public-private partnerships (PPPs) can also play a pivotal role in strengthening the Agri-tech ecosystem. By combining public sector resources with private sector expertise, these partnerships can address systemic challenges such as access to finance, infrastructure, and market linkages. For instance, a PPP initiative could focus on developing rural digital hubs equipped with internet access, training facilities, and demonstration farms to showcase the benefits of agricultural technologies to local communities.

4.2 Leveraging Data for Smart Agricultural Decision-Making

Data is the cornerstone of technology-driven agricultural supply chains, providing insights that enable smarter decision-making at every stage of the chain. In Nigeria, leveraging data effectively can help optimize resource allocation, enhance market coordination, and mitigate risks such as climate variability and supply chain disruptions. However, the collection, analysis, and application of agricultural data remain underutilized, presenting a significant opportunity for improvement.

One of the most promising applications of data in agriculture is precision farming, where real-time data from sensors, drones, and satellite imagery informs decisions about planting, irrigation, pest control, and harvesting. For example, soil moisture sensors can provide farmers with precise information on when and how much to irrigate, conserving water and improving crop yields. Similarly, drone-mounted cameras can monitor crop health, identify pest infestations early, and guide targeted interventions, reducing the need for broad-spectrum pesticide use.

Market intelligence platforms that aggregate data on prices, demand, and supply trends can also enhance the efficiency of agricultural supply chains. Farmers and traders can use these platforms to identify optimal times and locations for selling produce, reducing price volatility and improving profitability. Additionally, data analytics tools can help processors and retailers forecast demand patterns, enabling better inventory management and reducing food waste.

The integration of blockchain technology with data systems offers an added layer of transparency and accountability. By recording every transaction and movement of produce on a decentralized ledger, blockchain ensures traceability, a critical factor for export commodities and high-value crops. For instance, a blockchain-based system could verify the origin and quality of cocoa beans, meeting the requirements of international buyers and opening access to premium markets.

However, to fully leverage data for agricultural transformation, significant investments in digital infrastructure are required. Expanding internet connectivity in rural areas, deploying affordable data collection devices, and training stakeholders in data literacy are essential steps. Collaborations with technology firms and international development agencies can accelerate the deployment of these solutions, ensuring that even the most remote farming communities' benefit from data-driven agriculture.

CHAPTER 5

Policy Frameworks and Governance in Agricultural Supply Chains

Agricultural supply chains do not operate in isolation, they are shaped by a complex web of policies, regulations, and governance structures that determine their efficiency, inclusivity, and sustainability. In Nigeria, a supportive policy environment is essential for addressing structural challenges, incentivizing investments, and promoting the adoption of innovations that enhance supply chain performance. This chapter explores the critical role of policy frameworks and governance in transforming Nigeria's agricultural supply chains, identifying existing gaps, and proposing actionable solutions for improvement.

The Nigerian government has made several attempts to reform the agricultural sector through policies such as the Agricultural Transformation Agenda (ATA) and the Green Alternative Framework. These initiatives have sought to modernize farming practices, increase productivity, and improve market linkages. However, the success of these policies has been hindered by inconsistent implementation, inadequate funding, and weak institutional capacity. To achieve meaningful transformation, agricultural policies must be designed with a clear focus on supply chain efficiency and inclusivity, addressing bottlenecks in production, processing, transportation, and marketing.

Land tenure is one of the most significant policy challenges affecting Nigeria's agricultural supply chains. The current land tenure system, governed by a mix of statutory and customary laws, often leaves farmers with insecure land rights, discouraging long-term investments in productivity-enhancing technologies and infrastructure. Streamlining land policies to provide farmers with clear, transferable, and enforceable land rights is critical for unlocking the potential of agricultural supply chains. Policies that promote land consolidation, particularly for smallholder farmers, can also enable economies of scale and improve access to markets and financing.

Trade policies play a crucial role in shaping the competitiveness of Nigerian agricultural products in both domestic and international markets. Import restrictions and export bans, often introduced to protect local industries or stabilize prices, can have unintended consequences on supply chains. For example, restrictions on the export of certain crops may discourage farmers from producing surplus quantities, reducing the availability of raw materials for processors. Harmonizing trade policies to ensure consistency and predictability is essential for fostering investor confidence and encouraging private-sector participation in agricultural supply chains.

Subsidies and incentives are another critical area of policy intervention. While input subsidies, such as those for fertilizers and seeds, have been widely used to support farmers, their distribution is often marred by inefficiencies and corruption. A more effective approach would involve targeting subsidies to smallholder farmers through digital platforms, ensuring transparency and accountability. Additionally, providing incentives for investments in storage, transportation, and processing infrastructure can enhance supply chain efficiency and reduce post-harvest losses.

Governance structures also have a significant impact on the performance of agricultural supply chains. Weak institutions, fragmented coordination among stakeholders, and inadequate enforcement of regulations are common challenges in Nigeria. Strengthening governance requires establishing clear roles and responsibilities for government agencies, private-sector actors, and civil society organizations involved in agricultural supply chains. For instance, creating an independent regulatory body to oversee logistics and quality standards can ensure compliance and promote fair competition.

Public-private partnerships (PPPs) offer a promising model for addressing governance challenges and driving investment in agricultural supply chains. By leveraging private-sector expertise and resources, PPPs can bridge gaps in infrastructure, technology, and financing. For example, a PPP could focus on developing a network of cold storage facilities across major agricultural regions, reducing post-harvest losses and improving market access for farmers.

Policy coherence is another critical factor for success. Agricultural supply chains are influenced by policies from multiple sectors, including transportation, energy, trade, and environment. Ensuring that these policies align with agricultural objectives is essential for avoiding contradictions and maximizing impact. For example, transportation policies that prioritize the development of rural roads can complement agricultural policies aimed at improving market access, creating synergies that benefit the entire supply chain.

To deepen the impact of policy frameworks and governance on Nigeria's agricultural supply chains, it is essential to explore the interplay between various policy domains and their practical implications. Effective governance does not solely rely on the formulation of policies but also hinges on their execution, monitoring, and adaptation to changing realities. Expanding on these themes, this section will examine the need for stronger institutional capacity, stakeholder engagement, and accountability mechanisms to ensure the success of policy interventions.

One of the critical barriers to effective policy implementation in Nigeria is the limited capacity of institutions tasked with overseeing agricultural supply chains. Many government agencies lack the technical expertise, financial resources, and infrastructure required to carry out their mandates effectively. For example, while Nigeria has several policies aimed at reducing post-harvest losses, the absence of well-trained extension workers, monitoring systems, and enforcement mechanisms undermines these efforts. Strengthening institutional capacity requires targeted investments in human resources, infrastructure, and technology.

First, capacity-building programs should be implemented to train government officials, extension workers, and other stakeholders involved in policy implementation. These programs should focus on enhancing skills in areas such as data collection and analysis, supply chain management, and regulatory enforcement. Additionally, equipping institutions with modern technology, such as digital monitoring tools and data management systems, can improve efficiency and transparency in policy implementation.

Second, decentralizing certain aspects of governance can enhance responsiveness and accountability. Local governments, with their proximity to farming communities, are better positioned to address specific challenges within their regions. Empowering local governments with the authority, resources, and technical support needed to implement agricultural policies can improve outcomes. For instance, local governments could oversee the construction and maintenance of rural roads or manage cooperatives that aggregate produce for market access.

Third, fostering collaboration between public and private institutions can address capacity gaps. Private-sector companies often possess the expertise and resources needed to implement complex projects, such as the development of logistics hubs or digital platforms. By partnering with these companies, government agencies can leverage their capabilities while focusing on their core regulatory functions.

Stakeholder engagement is a cornerstone of effective governance. Agricultural supply chains are inherently complex, involving diverse actors such as farmers, processors, traders, transporters, and policymakers. Policies that are developed without the input of these stakeholders often fail to address their needs and priorities, leading to poor adoption and unintended consequences. Therefore, creating platforms for inclusive dialogue and collaboration is crucial for the success of policy interventions.

Regular consultations with farmer associations, cooperatives, and agribusiness representatives can ensure that policies are grounded in the realities of those directly affected. For example, engaging smallholder farmers in discussions about land tenure reforms can help policymakers design solutions that address local challenges, such as disputes over land boundaries or inheritance rights. Similarly, processors and exporters can provide valuable insights into trade policies, highlighting barriers that hinder competitiveness in international markets.

Transparency is another critical element of governance that fosters accountability and trust. Farmers and other supply chain actors must have access to information about policies, subsidies, and regulations that affect their livelihoods. Digital platforms can play a transformative role in this regard, providing real-time updates on government programs, market prices, and regulatory requirements. For example, an online portal that tracks the distribution of subsidized inputs, such as fertilizers and seeds, can reduce corruption and ensure that resources reach their intended beneficiaries.

Monitoring and evaluation (M&E) systems are essential for assessing the effectiveness of policies and identifying areas for improvement. Robust M&E frameworks should include measurable indicators, such as reductions in post-harvest losses, increases in market access, or improvements in farmer incomes. Independent audits and impact assessments can provide unbiased evaluations of policy outcomes, enabling policymakers to refine their approaches.

Finally, enforcing compliance with regulations is a critical aspect of governance that often goes overlooked. In Nigeria, weak enforcement mechanisms allow issues such as substandard inputs, poor transportation practices, and market monopolies to persist. Strengthening enforcement requires well-defined legal frameworks, adequately resourced regulatory bodies, and penalties for non-compliance. For example, setting clear standards for produce quality and implementing penalties for fraudulent practices can enhance the credibility of Nigeria's agricultural supply chains in both domestic and international markets.

5.1 Building Policy Synergies Across Sectors

Agricultural supply chains in Nigeria are influenced by policies from multiple sectors, including transportation, energy, education, and trade. For policies to be effective, they must be harmonized to create synergies that amplify their impact. Fragmented or contradictory policies often result in inefficiencies and missed opportunities for growth. For instance, an agricultural policy that encourages increased production may fail if transportation policies do not address the need for improved rural road networks to facilitate market access. Similarly, energy policies that do not prioritize rural electrification hinder the operation of cold storage facilities and processing plants essential for reducing post-harvest losses.

Harmonization begins with inter-ministerial coordination. Ministries of agriculture, transportation, energy, and trade must work collaboratively to design and implement policies that align with overarching goals for agricultural development. Establishing a central coordination body or task force can facilitate this process, ensuring that policies are complementary and not working at cross-purposes. For example, a policy to promote the export of agricultural goods could be linked with transportation initiatives to develop logistics hubs near major ports and energy strategies to power these facilities with renewable energy.

Another critical aspect of policy synergy is integrating agricultural goals into broader national development plans. For example, Nigeria's Economic Recovery and Growth Plan (ERGP) already recognizes agriculture as a key sector for economic diversification. Ensuring that policies at all levels; federal, state, and local, align with this vision can drive coherent action. State governments, in particular, play a pivotal role in addressing region-specific challenges, such as tailoring irrigation policies to arid regions or developing specialized processing facilities for locally dominant crops.

Public-private dialogue also enhances policy synergies by involving the private sector in decision-making processes. Private companies often have a clearer understanding of operational bottlenecks and can provide actionable recommendations for aligning policies with market realities. Regular forums for dialogue between policymakers and private sector stakeholders can foster mutual understanding and pave the way for collaborative solutions.

5.2 Incentivizing Private Sector Participation

The private sector is a critical driver of innovation, investment, and efficiency in agricultural supply chains. However, private sector participation in Nigeria's agriculture is often limited by perceived risks, regulatory uncertainty, and infrastructure deficits. Incentivizing private-sector involvement through targeted policies and strategic partnerships is essential for unlocking the full potential of Nigeria's agricultural sector.

Tax incentives are a proven strategy for encouraging private investment in agriculture. For example, providing tax breaks for companies that invest in processing facilities, cold storage units, or transportation networks can lower their financial risks and stimulate long-term investment. Similarly, tax holidays or reduced tariffs for importing agricultural equipment and technology can accelerate the adoption of modern practices and infrastructure development. These incentives should be clearly communicated and consistently applied to build investor confidence.

Access to credit is another critical area where policy interventions can incentivize private sector participation. While Nigeria has several agricultural financing programs, many private companies struggle to secure affordable loans due to high interest rates and stringent collateral requirements. Establishing credit guarantee schemes, where the government shares the risk with financial institutions, can encourage banks to lend to agribusinesses. Additionally, promoting blended finance models that combine public funding with private capital can mobilize resources for large-scale infrastructure projects, such as developing agro-industrial parks or logistics hubs.

Public-private partnerships (PPPs) are particularly effective for addressing infrastructure challenges. These partnerships leverage the expertise and efficiency of the private sector while ensuring that projects align with public priorities. For example, a PPP to construct a network of rural roads could involve the government providing land and regulatory approvals, while private companies finance, build, and maintain the infrastructure. PPPs can also facilitate technology transfer, as private firms often have access to advanced technologies that can improve supply chain efficiency.

Finally, creating an enabling business environment is essential for attracting private investment. This includes streamlining regulatory processes, reducing bureaucratic red tape, and ensuring the rule of law. A transparent and predictable regulatory framework not only reduces operational risks for private companies but also signals the government's commitment to fostering a thriving agricultural sector.

In conclusion, building policy synergies across sectors and incentivizing private sector participation are key strategies for transforming Nigeria's agricultural supply chains. These efforts must be underpinned by strong governance structures and a commitment to transparency, ensuring that policies deliver tangible benefits for all stakeholders. As the next chapter explores financing agricultural supply chains, it will build on these insights to identify innovative models and mechanisms for mobilizing resources on a scale.

CHAPTER 6

Financing Agricultural Supply Chains

Financing is the lifeblood of any functional and efficient supply chain. In Nigeria, where the agricultural sector is predominantly composed of smallholder farmers and micro- and small-sized enterprises, access to adequate and affordable financing remains a significant challenge. Without sufficient capital, farmers and other supply chain actors struggle to invest in productivity-enhancing technologies, infrastructure, and operations, limiting their ability to scale and contribute to the broader economy. This chapter explores the current state of financing in Nigeria's agricultural supply chains, highlights innovative financing models, and outlines strategies for improving access to financial resources.

Agricultural supply chains require financing at multiple levels, including production, processing, storage, transportation, and distribution. However, traditional financial institutions often perceive agriculture as a high-risk sector due to factors such as unpredictable weather patterns, fluctuating commodity prices, and limited collateral. As a result, farmers and small agribusinesses face high interest rates, stringent loan conditions, and limited access to formal credit. This financing gap perpetuates inefficiencies in the supply chain, leading to high post-harvest losses, underutilized processing capacities, and weak market linkages.

To bridge the financing gap in Nigeria's agricultural sector, innovative financing models have emerged as viable alternatives to traditional lending. These models leverage technology, partnerships, and risk-sharing mechanisms to provide tailored solutions for farmers and agribusinesses.

One promising model is value chain financing (VCF), which integrates financial services into the agricultural value chain. In a VCF system, buyers, processors, and other downstream actors provide credit or inputs to farmers, which are repaid through future sales. This approach aligns the interests of all stakeholders and reduces the risk of default, as payments are linked to the delivery of goods. For instance, a cocoa processor may supply farmers with seeds, fertilizers, and technical support in exchange for a guaranteed portion of their harvest.

Digital financial platforms are another transformative innovation. Mobile money services and fintech solutions enable farmers to access credit, savings, and insurance products through their mobile phones. Platforms like Farmcrowdy in Nigeria allow individuals to invest in farms, providing farmers with upfront capital while investors earn a return from the harvest. These platforms democratize access to finance and reduce transaction costs, particularly for smallholder farmers in remote areas.

Blended finance is an approach that combines public and private funding to de-risk investments in agriculture. By using public funds to provide guarantees, subsidies, or concessional loans, governments and development agencies can attract private capital to high-impact projects. For example, a blended finance initiative could support the development of cold storage facilities or logistics hubs, enabling supply chain actors to access affordable infrastructure.

Weather-indexed insurance is another innovative financial product that addresses the risks associated with climate variability. Instead of compensating farmers based on actual losses, this type of insurance pays out when predefined weather conditions, such as low rainfall or extreme temperatures, occur. This reduces administrative costs and provides farmers with a safety net, encouraging them to invest in their operations despite climate-related uncertainties.

Expanding access to finance requires a coordinated effort among governments, financial institutions, and development partners. One key strategy is strengthening the institutional capacity of agricultural development banks, such as the Bank of Agriculture (BOA) in Nigeria. By recapitalizing these banks, modernizing their operations, and enhancing their outreach, governments can ensure that smallholder farmers and agribusinesses have access to affordable credit.

Public-private partnerships (PPPs) can also play a critical role in mobilizing resources for agricultural financing. For instance, partnerships between banks and agribusinesses can create loan products tailored to the needs of supply chain actors. Governments can further incentivize these collaborations by offering tax breaks or co-financing arrangements.

Expanding financial literacy and inclusion is equally important. Many farmers in Nigeria lack the knowledge and skills needed to navigate formal financial systems, limiting their access to credit and insurance products. Financial literacy programs can educate farmers on budgeting, record-keeping, and loan management, empowering them to engage with banks and fintech platforms. Additionally, promoting the use of mobile money and digital wallets can increase financial inclusion, particularly in underserved rural areas.

The government can also improve access to finance by creating a conducive policy environment. This includes streamlining regulations for microfinance institutions, reducing bureaucratic barriers, and ensuring the enforceability of contracts. Establishing credit guarantee schemes, where the government shares the risk of lending with financial institutions, can further encourage banks to extend credit to farmers and agribusinesses.

While innovative financing models and strategic interventions address immediate gaps, creating a sustainable financial ecosystem for Nigeria's agricultural supply chains requires long-term structural changes. These changes must focus on fostering collaboration among stakeholders, improving financial infrastructure, and aligning financing solutions with the specific needs of supply chain actors.

One critical step is the development of robust financial ecosystems that link farmers, agribusinesses, financial institutions, and policymakers. This ecosystem should include specialized financial products tailored to different segments of the supply chain. For instance, short-term credit facilities can help farmers purchase input during the planting season, while long-term loans can support investments in infrastructure, such as cold storage units and processing facilities. Agribusinesses may benefit from working capital loans to manage cash flow and meet operational expenses.

Agricultural cooperatives can play a pivotal role in strengthening these ecosystems by serving as intermediaries between farmers and financial institutions. By pooling resources, cooperatives reduce the risks associated with lending to individual farmers, making it more attractive for banks to extend credit. Additionally, cooperatives can negotiate better terms for loans and other financial products, ensuring that farmers benefit from lower interest rates and more favorable repayment schedules.

Another key area of focus is the expansion of financial infrastructure in rural areas. Many farming communities in Nigeria lack access to basic banking services, making it difficult for farmers to save, borrow, or transfer money. Expanding the reach of microfinance institutions, mobile banking agents, and rural banking branches is essential for bridging this gap. Governments and development agencies can support these efforts by providing grants, technical assistance, and regulatory incentives to financial institutions willing to establish operations in underserved areas.

The integration of digital technology into financial ecosystems can further enhance efficiency and accessibility. Digital platforms that aggregate financial data, such as credit histories and repayment patterns, can help lenders assess the creditworthiness of farmers and small agribusinesses more accurately. This reduces the perceived risks of lending and allows financial institutions to extend credit to previously underserved populations. Additionally, mobile money platforms and digital wallets enable seamless transactions, reducing reliance on cash and improving financial security.

Agriculture in Nigeria is often perceived as a high-risk sector due to factors such as climate variability, price volatility, and market fragmentation. De-risking investments in agricultural supply chains is crucial for attracting private capital and fostering a culture of sustainable financing. Governments, development agencies, and financial institutions must work together to design mechanisms that mitigate these risks and create a more favorable investment climate.

One effective de-risking strategy is the establishment of credit guarantee schemes. Under these schemes, the government or a development agency shares the risk of lending with financial institutions, covering a portion of potential loan defaults. This reduces the financial exposure of banks and encourages them to extend credit to farmers and agribusinesses. For example, a credit guarantee scheme could target smallholder farmers engaged in high-value crops, such as cocoa or sesame, providing them with the capital needed to scale their operations and access premium markets.

Insurance products are another critical tool for de-risking investments. Weather-indexed insurance, in particular, provides a safety net for farmers facing climate-related risks, such as droughts or floods. By linking payouts to specific weather conditions, this type of insurance eliminates the need for costly and time-consuming claims assessments. Governments and development agencies can subsidize premiums for smallholder farmers, ensuring that these products are affordable and accessible.

Risk-sharing facilities can also facilitate investment in infrastructure and technology. For instance, a risk-sharing facility for cold storage development could provide partial guarantees to investors, reducing the financial risks associated with constructing and operating these facilities. Similarly, co-financing arrangements between public and private entities can de-risk investments in digital platforms, logistics systems, and processing hubs.

Policy reforms are essential for creating an environment that supports de-risking. Clear and enforceable land tenure policies, stable trade regulations, and streamlined permitting processes reduce uncertainties for investors. Additionally, promoting transparency and accountability in government programs builds trust among stakeholders, encouraging greater private sector participation.

6.1 Fostering Inclusive Financing Mechanisms

Inclusivity in agricultural financing is paramount to ensuring that all stakeholders, particularly smallholder farmers, women, and youth, can access the resources needed to thrive in Nigeria's agricultural supply chains. Historically, these groups have faced systemic barriers to finance due to limited collateral, lack of financial literacy, and cultural or institutional biases. Addressing these disparities is not just a matter of equity but also a strategic imperative for unlocking the full potential of Nigeria's agricultural sector.

Smallholder farmers, who constitute the majority of Nigeria's agricultural workforce, often lack the formal documentation or collateral needed to secure traditional loans. To bridge this gap, financial institutions can adopt alternative credit assessment methods, such as leveraging transaction histories from mobile money platforms or data from cooperative organizations. Digital credit scoring systems, powered by machine learning algorithms, can evaluate a farmer's creditworthiness based on non-traditional metrics like input purchase records, harvest data, and repayment behavior on smaller loans.

Targeted financial products designed specifically for women and youth can also enhance inclusivity. Women, who make up a significant portion of Nigeria's agricultural labor force, often face cultural and legal barriers to owning land or accessing credit. Microfinance programs tailored for women, coupled with capacity-building initiatives, can empower them to invest in their farms and businesses. Similarly, youth-focused financing schemes, such as agribusiness start-up grants or subsidized loans for young entrepreneurs, can attract the next generation to agriculture and foster innovation in the sector.

Partnerships with NGOs, development agencies, and private foundations can amplify the impact of inclusive financing initiatives. These organizations can provide technical assistance, grants, or capacity-building programs to complement financial products. For instance, an NGO might offer financial literacy training to women farmers, equipping them with the skills to manage loans and invest wisely. Similarly, international development agencies can co-finance programs targeting underserved groups, reducing the risk for local financial institutions and enabling broader outreach.

6.2 Aligning Financing with Sustainable Development Goals (SDGs)

The alignment of agricultural financing with the United Nations' Sustainable Development Goals (SDGs) offers an opportunity to integrate social, economic, and environmental objectives into Nigeria's agricultural supply chains. Financing models that prioritize sustainability not only address immediate challenges but also contribute to long-term resilience and global development priorities.

One critical area for alignment is financing climate-smart agriculture (CSA). Climate-smart practices, such as conservation tillage, agroforestry, and efficient irrigation systems, require upfront investments that many farmers cannot afford. By providing concessional loans, grants, or subsidies for CSA technologies, financial institutions can incentivize the adoption of sustainable practices. Additionally, green bonds and climate funds can mobilize resources for large-scale projects, such as renewable energy-powered cold storage facilities or reforestation initiatives.

Socially responsible investment (SRI) funds represent another avenue for aligning financing with the SDGs. These funds prioritize projects that deliver social and environmental benefits alongside financial returns. For example, an SRI fund might invest in a supply chain initiative that provides fair trade certification for cocoa farmers, ensuring better wages and working conditions while enhancing export potential. Encouraging local and international investors to channel their resources into such funds can scale impact across Nigeria's agricultural sector.

Impact investing is a growing trend that aligns perfectly with the SDGs. Investors in this space seek measurable social or environmental outcomes in addition to financial returns. For instance, an impact investment fund might support a fintech company developing a mobile app that connects smallholder farmers with microloans and technical advisory services. These investments not only drive innovation but also enhance the resilience and productivity of supply chains.

Policymakers have a critical role in creating an enabling environment for sustainable financing. This includes introducing incentives for financial institutions to adopt sustainability criteria in their lending practices, such as lower tax rates for green or social-impact loans. Additionally, governments can establish sustainability standards for agricultural projects seeking public funding or guarantees, ensuring that financed activities align with national and global development goals.

In conclusion, fostering inclusive financing mechanisms and aligning them with the SDGs are essential for creating a resilient, equitable, and sustainable agricultural sector in Nigeria. These efforts not only address the immediate needs of farmers and supply chain actors but also position the sector as a key driver of national development and global sustainability. As the focus shifts to market access and export potential in the next chapter, these financing foundations will play a critical role in enabling Nigerian agriculture to compete and thrive in domestic and international markets.

CHAPTER 7

Market Access and Export Potential

Market access and export potential are critical components of a robust agricultural supply chain. In Nigeria, despite being Africa's largest economy, agricultural products often struggle to reach their full market value due to inefficiencies, lack of infrastructure, and limited competitiveness in global trade. Enhancing market access and capitalizing on export opportunities can significantly increase the incomes of farmers, reduce post-harvest losses, and position Nigeria as a major player in international agricultural markets. This chapter explores the challenges and opportunities in improving market access and export potential, with a focus on strategies for integrating Nigerian agriculture into domestic, regional, and global value chains.

The first step in unlocking the full potential of Nigeria's agricultural sector lies in strengthening domestic market linkages. With a population exceeding 200 million, Nigeria represents a significant consumer market for agricultural products. However, inefficiencies in logistics, fragmented supply chains, and poor market infrastructure often prevent farmers from accessing urban markets where demand and prices are higher.

One of the most pressing challenges is the lack of reliable transportation networks. Farmers in rural areas often face exorbitant transportation costs and delays due to poorly maintained roads. Investment in rural road construction and maintenance is essential for improving market access. Additionally, developing multimodal transportation systems, such as combining road, rail, and waterway networks, can reduce costs and enhance efficiency.

Market infrastructure also requires significant upgrades. Traditional markets often lack basic facilities, such as storage, grading, and weighing systems, which are essential for ensuring product quality and fair pricing. Establishing modern wholesale markets and aggregation centers can provide farmers with access to larger buyers, including processors, retailers, and exporters. These facilities can also serve as hubs for quality control and certification, increasing the marketability of Nigerian agricultural products.

Digital platforms are another game-changer for improving domestic market access. Mobile applications and e-commerce platforms can connect farmers directly to buyers, eliminating intermediaries and ensuring fairer prices. For example, platforms like AFEX and Farmcrowdy in Nigeria have successfully linked farmers to domestic markets, providing them with real-time price information and guaranteed offtake agreements.

While domestic markets offer substantial opportunities, tapping into regional and global markets is essential for driving the growth and competitiveness of Nigeria's agricultural sector. Exporting agricultural products not only generates foreign exchange but also incentivizes farmers and processors to adopt higher standards and technologies.

Nigeria's membership in regional trade agreements, such as the African Continental Free Trade Area (AfCFTA), presents significant opportunities for expanding market access across Africa. By eliminating tariffs and non-tariff barriers, the AfCFTA creates a single market of over 1.3 billion people, offering Nigerian farmers and agribusinesses access to new customers and supply chain partnerships. For instance, Nigerian cassava products, such as flour and ethanol, can find ready markets in neighboring countries where demand exceeds domestic production.

However, capitalizing on these opportunities requires addressing barriers to competitiveness. Nigerian agricultural exports often face challenges related to quality, certification, and traceability. For example, exports of cocoa and sesame seeds are sometimes rejected in international markets due to contamination or failure to meet phytosanitary standards. Strengthening quality control systems and adopting international certifications, such as Global GAP or Fair Trade, can enhance the reputation of Nigerian products and open access to premium markets.

Trade facilitation measures are also critical. Simplifying customs procedures, reducing port congestion, and implementing digital trade platforms can significantly reduce the cost and time associated with exporting agricultural goods. Investments in export logistics, such as cold storage at ports and specialized export terminals, further enhance the efficiency of agricultural supply chains.

Value addition is another key strategy for maximizing export potential. Nigeria currently exports many of its agricultural products in raw form, missing out on the higher revenues associated with processed goods. For example, instead of exporting raw cocoa beans, Nigeria could produce and export chocolate and cocoa butter, capturing more value within the country. Government incentives, such as tax breaks for processing facilities and grants for research and development, can encourage investments in value addition.

Finally, building strategic trade partnerships with countries and regions outside Africa can diversify Nigeria's export markets. Bilateral trade agreements with the European Union, Middle Eastern countries, or Asia-Pacific nations can create dedicated market access channels for Nigerian products. Trade promotion initiatives, such as participation in international trade fairs and marketing campaigns, can further increase the visibility and competitiveness of Nigerian agriculture on the global stage.

Regional trade offers a substantial growth opportunity for Nigeria's agricultural sector, particularly within West Africa. Proximity to neighboring countries provides logistical advantages, while shared cultural and dietary preferences create demand for similar agricultural products. For example, grains like maize and sorghum, tubers such as yam and cassava, and livestock products are in high demand across the

Economic Community of West African States (ECOWAS). Strengthening cross-border trade within this region can enhance Nigeria's agricultural exports, providing farmers and agribusinesses with new revenue streams.

One key area for improvement is the harmonization of regional trade policies. Despite Nigeria's membership in ECOWAS and the African Continental Free Trade Area (AfCFTA), inconsistent tariffs, border closures, and non-tariff barriers often disrupt trade. Establishing a unified framework for agricultural trade across the region can reduce these obstacles. For example, implementing standardized quality and safety regulations ensures that Nigerian goods meet the requirements of importing countries, streamlining the movement of goods.

Logistics and border infrastructure are also critical for facilitating regional trade. Many of Nigeria's border crossings lack modern facilities, leading to delays, corruption, and increased transaction costs. Investments in customs modernization, such as the introduction of electronic documentation systems and single-window platforms, can expedite clearance processes. Additionally, developing export corridors with well-maintained roads, aggregation centers, and storage facilities can create seamless connections between production areas and regional markets.

Regional value chains can also be strengthened through partnerships and collaboration among countries. For instance, Nigeria could collaborate with neighboring countries to create integrated supply chains for specific commodities. In the case of rice, Nigeria could focus on production while partnering with countries that have advanced milling and packaging capabilities. Such partnerships distribute value addition activities across the region, creating shared economic benefits and fostering regional integration.

Finally, addressing informal trade is crucial. A significant portion of cross-border agricultural trade in West Africa occurs informally, bypassing official channels and depriving governments of revenue. Policies that simplify trade formalization, such as reducing fees and providing incentives for small-scale traders to register, can bring informal trade into the formal economy, increasing transparency and accountability.

For Nigeria to compete in global agricultural markets, it must invest in export-ready infrastructure and build ecosystems that support international trade. This involves not only enhancing physical infrastructure but also creating institutional frameworks and support systems that meet the demands of global buyers.

Modern export logistics are a cornerstone of this effort. Nigeria's ports, often plagued by congestion and inefficiency, require significant upgrades to handle the increasing volume of agricultural exports. Establishing dedicated agricultural export terminals equipped with cold storage, packaging facilities, and quality control laboratories can ensure that perishable goods maintain their freshness and comply with international standards. Expanding access to inland dry ports can further reduce the logistical burden on coastal facilities, allowing goods to be processed and certified closer to production zones before being transported to seaports.

Air freight services are another untapped opportunity for high-value agricultural products, such as fresh fruits, vegetables, and flowers. Developing agro-cargo airports in key agricultural regions can facilitate the export of these products to international markets where demand for fresh, high-quality produce is strong. For example, Nigeria's mangoes and pineapples could gain traction in Middle Eastern and European markets with reliable air freight connections.

In addition to physical infrastructure, export readiness requires robust quality assurance systems. Strengthening institutions responsible for food safety, phytosanitary standards, and export certification is essential. Investments in laboratories, training for inspectors, and digital tracking systems can help Nigeria meet the stringent requirements of international buyers. For example, a blockchain-based traceability system could verify the journey of cocoa beans from Nigerian farms to European chocolatiers, ensuring compliance with ethical sourcing and quality standards.

Building export ecosystems also involves supporting smallholder farmers and processors to meet international market requirements. Export readiness programs can provide technical assistance on meeting standards, such as residue testing for pesticides or certifications like Global GAP and organic labeling. Additionally, financial incentives such as grants or subsidies for exporters investing in compliance measures can encourage broader participation in international markets.

Market diversification is another critical strategy. While Nigeria's agricultural exports have traditionally focused on a narrow range of commodities, such as cocoa and sesame seeds, expanding into new products and markets can reduce vulnerability to price fluctuations and demand shifts. Emerging markets in Asia, the Middle East, and Latin America present opportunities for Nigerian products, including processed goods like cassava flour, spices, and palm oil derivatives.

Strengthening regional trade and investing in export-ready infrastructure are pivotal for unlocking Nigeria's agricultural export potential. By enhancing cross-border trade, streamlining logistics, and building ecosystems that support compliance with international standards, Nigeria can position itself as a major player in global agricultural markets. These efforts not only benefit farmers and agribusinesses but also contribute to broader economic diversification and resilience.

7.1 Expanding Agricultural Export Capacity Through Value Addition

One of the most transformative strategies for enhancing Nigeria's agricultural export potential is the focus on value addition. Exporting raw agricultural commodities limits the economic benefits to the country, as the most significant profits are often realized during processing and packaging stages, which take place abroad. By shifting focus toward processing agricultural products domestically, Nigeria can capture more value within its borders, create jobs, and boost foreign exchange earnings.

Value addition begins with investment in processing infrastructure. Facilities for milling grains, extracting oils, fermenting cocoa, and packaging fresh produce are essential for transforming raw materials into finished or semi-finished goods that command higher prices on the global market. For instance, Nigeria is a leading producer of cocoa but primarily exports raw beans. Establishing cocoa processing plants to produce chocolate, cocoa powder, or cocoa butter could significantly increase export revenues while reducing reliance on raw commodity markets.

Government incentives play a pivotal role in encouraging value addition. Tax breaks, subsidies, or grants for agribusinesses investing in processing infrastructure can lower the cost of entry and accelerate the establishment of processing facilities. For example, a subsidy on equipment for rice milling or fruit canning plants can encourage private sector participation, driving investment in high-value segments of the supply chain. Additionally, public-private partnerships (PPPs) can facilitate the construction of agro-industrial parks equipped with shared facilities like energy, water, and storage.

Marketing and branding are also critical components of value addition. Nigerian agricultural exports often lack the brand recognition that distinguishes products from competitors. Investing in marketing campaigns that promote Nigerian products as premium, ethically sourced, or organic can attract niche international markets.

Geographical Indications (GIs), such as identifying Nigerian yams or shea butter as unique due to their origin, can also enhance their appeal and fetch premium prices globally.

Finally, capacity-building initiatives for farmers and processors are essential. Training programs focused on processing techniques, quality assurance, and international certification standards ensure that value-added products meet the requirements of global buyers. By aligning local production practices with international demands, Nigeria can position itself as a reliable source of high-quality, processed agricultural goods.

7.2 Addressing Supply Chain Inefficiencies to Boost Export Competitiveness

Supply chain inefficiencies are a significant barrier to Nigeria's competitiveness in global agricultural markets. High transportation costs, delays at ports, and inadequate storage facilities often result in lower-quality exports, reduced profit margins, and missed market opportunities. Addressing these inefficiencies is crucial for enhancing Nigeria's ability to compete with other agricultural exporters.

One of the primary challenges is the cost and reliability of transportation. Moving goods from rural production areas to export terminals is often expensive and time-consuming due to poor road networks and underdeveloped rail systems. Investments in rural roads and railways that connect agricultural hubs to ports can significantly reduce logistics costs and improve the timeliness of deliveries.

Additionally, optimizing inland waterway transportation, such as using barges to move goods along the Niger and Benue rivers, can provide a cost-effective alternative for bulky commodities like grains and oilseeds.

Port inefficiencies also hinder Nigeria's export capacity. Congestion at major ports, such as Apapa in Lagos, often leads to delays and increased costs for exporters. Streamlining customs procedures, implementing digital documentation systems, and expanding port capacity through dedicated agricultural export terminals are critical steps to address these challenges. For instance, introducing single-window platforms for export documentation can reduce clearance times and improve overall efficiency.

Storage and handling facilities at ports and along the supply chain are equally important. Many Nigerian agricultural products, particularly perishables, suffer from quality deterioration due to inadequate cold storage and warehousing. Developing modern cold chain infrastructure, including refrigerated trucks and storage units, can preserve the quality of perishable goods like fruits, vegetables, and fish, ensuring they meet international market standards.

Furthermore, improving coordination and communication across the supply chain can reduce bottlenecks and enhance efficiency. Digital platforms that connect farmers, transporters, processors, and exporters in real-time can ensure better planning and resource allocation. For example, a mobile app that provides real-time updates on transportation availability or port conditions can help exporters avoid delays and optimize their operations.

In conclusion, focusing on value addition and addressing supply chain inefficiencies are two critical pathways for unlocking Nigeria's agricultural export potential. These strategies not only enhance competitiveness but also ensure that Nigeria's agricultural sector delivers maximum economic benefits. The next chapter will examine sustainability in agricultural supply chains, emphasizing the importance of balancing growth with environmental and social responsibility.

CHAPTER 8

Sustainability in Agricultural Supply Chains

Sustainability is no longer a peripheral consideration in agriculture; it is a fundamental requirement for the long-term success of supply chains. For Nigeria, where agriculture forms the backbone of the economy and supports millions of livelihoods, integrating sustainable practices into supply chains is essential. This chapter explores the environmental, social, and economic dimensions of sustainability in agricultural supply chains and presents actionable strategies to balance growth with environmental stewardship and social equity.

Nigeria's agricultural sector heavily relies on natural resources such as soil, water, and forests. However, unsustainable practices, including overgrazing, deforestation, and the excessive use of chemical inputs, have degraded these resources, threatening the sector's long-term viability. To achieve environmental sustainability, supply chains must prioritize practices that conserve resources and reduce their ecological footprint.

One key approach is the adoption of climate-smart agriculture (CSA). CSA emphasizes practices that increase productivity while enhancing resilience to climate change and reducing greenhouse gas emissions. Examples include conservation tillage, which minimizes soil disturbance and preserves soil health, and agroforestry systems, which integrate trees into farmland to improve biodiversity and sequester carbon. Scaling CSA requires training programs, access to climate-resilient crop varieties, and financial incentives for farmers to adopt sustainable practices.

Water resource management is another critical area. Many Nigerian farmers rely on rain-fed agriculture, making them vulnerable to erratic weather patterns. Expanding irrigation infrastructure and promoting efficient water use technologies, such as drip irrigation and rainwater harvesting, can enhance productivity while conserving water. Solar-powered irrigation systems, in particular, offer a sustainable alternative for smallholder farmers with limited access to energy.

The reduction of waste is equally vital for environmental sustainability. Post-harvest losses, which account for up to 40% of Nigeria's agricultural output, represent not only economic inefficiencies but also an environmental burden. Investments in cold chain infrastructure, improved packaging, and processing facilities can significantly reduce waste. Additionally, recycling agricultural byproducts, such as turning crop residues into bioenergy or compost, supports a circular economy and reduces waste disposal challenges.

Governments and policymakers play a crucial role in driving environmental sustainability. Introducing regulations to limit deforestation, incentivizing organic farming practices, and establishing carbon credit systems can encourage farmers and agribusinesses to adopt greener practices. By integrating sustainability into agricultural policies, Nigeria can ensure that its growth aligns with global environmental goals.

Sustainability extends beyond environmental considerations to include social and economic dimensions. An inclusive supply chain is one that empowers marginalized groups, such as smallholder farmers, women, and youth, while ensuring fair distribution of economic benefits. For Nigeria, where inequality and rural poverty remain pervasive, promoting social and economic sustainability is critical for long-term development.

Smallholder farmers, who produce the majority of Nigeria's agricultural output, often face systemic barriers to accessing markets, finance, and resources. Strengthening farmer cooperatives and producer organizations can empower these farmers by providing them with collective bargaining power, access to shared resources, and opportunities for capacity building. For example, cooperatives can pool resources to invest in storage facilities or negotiate better prices with buyers.

Women, who constitute a significant portion of Nigeria's agricultural workforce, often face gender-based barriers, including limited access to land and financial services. Promoting gender equity in supply chains involves providing targeted support to women farmers, such as grants for purchasing inputs, training programs, and legal reforms to improve land ownership rights. Additionally, encouraging women's participation in leadership roles within cooperatives and agribusinesses ensures that their voices are heard in decision-making processes.

Youth inclusion is another critical aspect of social sustainability. With a growing youth population and high unemployment rates, agriculture offers a viable pathway for job creation. Programs that provide training in modern farming techniques, agribusiness management, and digital agriculture can attract young people to the sector. Additionally, youth-focused financing schemes, such as start-up grants for agribusiness ventures, can empower young entrepreneurs to innovate and drive change in supply chains.

Fair trade practices also contribute to social and economic sustainability. By ensuring that farmers and workers receive fair prices and wages, fair trade certifications can reduce exploitation and improve living standards. For example, cocoa and coffee farmers who participate in fair trade schemes often experience better income stability and improved access to global markets.

Technology plays a pivotal role in driving sustainability across agricultural supply chains. By integrating innovative tools and digital solutions, Nigeria can enhance resource efficiency, reduce environmental impacts, and ensure equitable distribution of benefits. These advancements are particularly relevant in addressing challenges such as resource management, traceability, and social inclusion.

Precision agriculture is one of the most promising technologies for sustainable farming. Using tools such as drones, satellite imaging, and soil sensors, farmers can optimize input usage, applying fertilizers, water, and pesticides only where and when they are needed. This targeted approach minimizes waste, reduces environmental pollution, and increases yields. For example, soil sensors can identify nutrient deficiencies in specific areas of a field, allowing farmers to apply precise quantities of fertilizers, thereby preserving soil health while reducing costs.

Blockchain technology offers significant potential for enhancing sustainability through traceability and transparency. By recording every transaction and movement of goods on an immutable ledger, blockchain can ensure that products are sourced and produced ethically and sustainably. This is particularly important for export commodities such as cocoa and coffee, where international buyers increasingly demand proof of ethical sourcing and adherence to environmental standards. Blockchain also strengthens trust among stakeholders by providing verifiable data on product origins, quality, and sustainability practices.

Mobile applications and digital platforms are transforming the way farmers access resources and markets. For example, apps that provide weather forecasts, pest management advice, or market price information enable farmers to make informed decisions that optimize resource use and reduce losses. Platforms that connect farmers directly with buyers or logistics providers can eliminate intermediaries, ensuring fairer prices and reducing inefficiencies in the supply chain.

Renewable energy technologies also play a vital role in sustainable supply chains. Solar-powered irrigation systems, cold storage units, and processing facilities reduce reliance on fossil fuels and lower carbon emissions. For example, solar-powered cold storage can extend the shelf life of perishable products, reducing post-harvest losses and ensuring that goods reach markets in optimal condition.

Policy frameworks are critical for embedding sustainability into Nigeria's agricultural supply chains. These frameworks must incentivize the adoption of sustainable practices while discouraging activities that deplete natural resources or perpetuate social inequalities. Governments, in collaboration with stakeholders, can create an enabling environment that drives sustainable growth.

Subsidies and tax incentives are effective tools for promoting sustainability. For example, subsidies for adopting climate-smart technologies, such as drip irrigation or renewable energy systems, can lower the financial barriers for farmers and agribusinesses. Similarly, tax breaks for companies that invest in sustainable processing facilities or fair-trade certification programs can encourage private-sector participation in sustainability initiatives.

Regulations that enforce environmental and social standards are equally important. For instance, policies that limit deforestation or mandate the use of sustainable farming practices can help preserve ecosystems and biodiversity. Establishing mandatory standards for waste management, such as recycling agricultural byproducts or reducing plastic use in packaging, ensures that supply chains operate with minimal environmental impact. To ensure compliance, governments can establish monitoring and enforcement mechanisms, supported by digital technologies like remote sensing and geospatial analysis.

Public-private partnerships (PPPs) can amplify the impact of policy frameworks by mobilizing resources and expertise from multiple sectors. For example, a PPP initiative could focus on developing an integrated network of renewable energy-powered cold storage facilities, combining public funding with private-sector innovation. Similarly, partnerships with NGOs and international development agencies can support the rollout of training programs and awareness campaigns on sustainability practices.

Education and awareness are also critical for fostering a culture of sustainability. Farmers and supply chain actors must understand the benefits of adopting sustainable practices, not only for the environment but also for their livelihoods. Training programs, workshops, and community engagement initiatives can build this awareness, empowering stakeholders to embrace sustainability as a core principle of their operations.

8.1 Strengthening Stakeholder Collaboration for Sustainable Practices

Achieving sustainability in Nigeria's agricultural supply chains requires coordinated efforts from all stakeholders, including farmers, agribusinesses, policymakers, researchers, and development partners. Collaboration not only fosters the sharing of knowledge and resources but also ensures that sustainability efforts are inclusive and aligned with the needs of all actors in the supply chain.

Farmers, as the primary stewards of agricultural resources, must be central to sustainability initiatives. Engaging farmers through cooperatives and producer organizations provides a platform for knowledge exchange, training, and collective action. For example, cooperatives can facilitate the adoption of sustainable practices by providing access to climate-resilient seeds, organic fertilizers, and shared equipment for conservation tillage. These organizations also give farmers a collective voice to advocate for policies that support sustainability.

Agribusinesses and processors are equally important in driving sustainability. Companies that source raw materials from Nigerian farmers can promote sustainable practices by offering incentives for compliance with environmental and social standards. For instance, a cocoa processor might pay a premium for beans produced using agroforestry methods, encouraging farmers to adopt these practices. Collaborating with certification bodies to implement fair trade and organic standards ensures that Nigerian agricultural products are competitive in international markets while promoting ethical and sustainable production.

Research institutions and universities play a pivotal role in generating knowledge and innovations that advance sustainability. Developing crop varieties that are high-yielding, drought-resistant, and pest-tolerant helps farmers adapt to changing climatic conditions. Additionally, research on efficient water use, renewable energy integration, and waste recycling provides practical solutions for reducing the environmental impact of agricultural supply chains. Partnerships between research institutions and private companies can accelerate the commercialization of these innovations, ensuring that they reach farmers and supply chain actors.

International development agencies and NGOs bring valuable expertise and funding to sustainability initiatives. These organizations can support capacity-building programs, pilot projects, and community-based initiatives that demonstrate the feasibility and benefits of sustainable practices. For example, an NGO might collaborate with local governments to implement a community-based irrigation project, combining technical training with infrastructure development to enhance water resource management.

8.2 Monitoring and Evaluating Sustainability Outcomes

Ensuring the long-term success of sustainability initiatives requires robust systems for monitoring and evaluating (M&E) their outcomes. M&E frameworks provide critical data on the effectiveness of interventions, enabling stakeholders to refine their strategies and scale successful approaches.

Developing sustainability indicators is the first step in creating an effective M&E framework. These indicators should capture the environmental, social, and economic dimensions of sustainability. For example, environmental indicators might include metrics such as reductions in greenhouse gas emissions, increases in soil organic matter, or improvements in water-use efficiency. Social indicators could track the participation of women and youth in supply chain activities, while economic indicators might measure changes in farmer incomes or export revenues.

Digital tools and technologies can enhance the accuracy and efficiency of M&E processes. Remote sensing and satellite imagery can monitor changes in land use, forest cover, and crop health over time, providing valuable insights into the environmental impact of agricultural practices. Mobile data collection tools allow fieldworkers to gather real-time information on farmer adoption of sustainable practices, crop yields, and supply chain activities. Blockchain technology can provide transparent records of compliance with sustainability standards, ensuring accountability among supply chain actors.

Stakeholder participation is essential for effective M&E. Farmers, agribusinesses, and community leaders should be involved in defining indicators, setting goals, and interpreting data. This participatory approach ensures that M&E frameworks reflect the realities on the ground and are aligned with local priorities. Additionally, involving stakeholders in the evaluation process builds trust and fosters a sense of ownership, increasing the likelihood of sustained engagement in sustainability efforts.

The results of M&E processes should be widely disseminated to inform decision-making and drive continuous improvement. Governments can use these findings to refine policies and allocate resources more effectively, while private companies can identify opportunities to enhance their supply chain operations. Public reports and case studies showcasing successful sustainability initiatives can inspire other stakeholders to adopt similar practices, creating a ripple effect across the agricultural sector.

In conclusion, strengthening stakeholder collaboration and implementing robust M&E frameworks are critical for embedding sustainability into Nigeria's agricultural supply chains. These efforts ensure that sustainability initiatives are inclusive, effective, and scalable, laying the foundation for a resilient and competitive agricultural sector. The next chapter will focus on addressing post-harvest losses, a major challenge in optimizing supply chain efficiency and achieving sustainability goals.

CHAPTER 9

Addressing Post-Harvest Losses in Agricultural Supply Chains

Post-harvest losses represent a significant challenge in Nigeria's agricultural supply chains, with estimates suggesting that up to 40% of agricultural produce is lost before reaching consumers. These losses occur at various stages of the supply chain, including harvesting, storage, transportation, and processing. The consequences of post-harvest losses extend beyond economic inefficiencies, affecting food security, farmer incomes, and environmental sustainability. This chapter explores the causes of post-harvest losses, their impact, and actionable solutions for reducing them to enhance the efficiency and profitability of Nigeria's agricultural sector.

Post-harvest losses in Nigeria are driven by a combination of inadequate infrastructure, lack of knowledge and technology, and systemic inefficiencies in the supply chain. Poor harvesting techniques, such as using inappropriate tools or harvesting at the wrong time, often result in physical damage to crops, making them more susceptible to spoilage. Additionally, improper handling during aggregation and transportation exacerbates these issues, especially for perishable goods like fruits, vegetables, and dairy products.

Storage is one of the most critical points where losses occur. A lack of adequate storage facilities, particularly in rural areas, forces farmers to sell their produce immediately after harvest, often at lower prices due to market gluts. Inadequate storage also leads to spoilage caused by pests, mold, and temperature fluctuations. Grains, for instance, are frequently damaged by weevils or fungal infections due to improper storage conditions.

Transportation challenges further compound post-harvest losses. Poor road networks, unreliable vehicles, and long transit times result in delays and damage to production. For example, tomatoes harvested in northern Nigeria often spoil before reaching southern markets due to the absence of refrigerated transportation. The absence of cold chain logistics disproportionately affects high-value perishable products, limiting their marketability and reducing farmer earnings.

The economic and environmental impacts of post-harvest losses are profound. For farmers, losses translate into reduced income and missed opportunities to reinvest in their operations. At the national level, these losses undermine food security by reducing the availability of nutritious food. Environmentally, post-harvest losses contribute to greenhouse gas emissions, as decomposing organic matter releases methane, a potent greenhouse gas. Additionally, the wasted resources used to produce lost food, such as water, fertilizers, and energy, exacerbate the environmental toll.

Addressing post-harvest losses requires a multi-pronged approach that combines infrastructure development, capacity building, technological innovation, and policy support. By targeting interventions at each stage of the supply chain, stakeholders can significantly reduce losses and improve overall efficiency.

Investing in storage infrastructure is one of the most effective ways to reduce post-harvest losses. Developing modern storage facilities, such as silos for grains and cold storage units for perishable goods, ensures that produce can be preserved for longer periods without spoilage. For smallholder farmers, community-based storage solutions, such as cooperatively owned warehouses, provide an affordable alternative to individual investments. Portable storage technologies, like hermetic bags, are also valuable for preserving grain quality by preventing pest infestations and moisture ingress.

Improving transportation logistics is another critical area of intervention. Upgrading rural roads and expanding rail networks can reduce transit times and minimize damage to produce during transportation. Refrigerated vehicles and cold chain logistics are particularly important for maintaining the quality of perishable goods, enabling farmers to access distant markets without losses. Public-private partnerships (PPPs) can play a vital role in financing and developing these logistics systems.

Capacity building and training programs are essential for equipping farmers and supply chain actors with the knowledge and skills needed to handle produce effectively. Training in proper harvesting techniques, grading, and packaging can reduce physical damage and improve the marketability of agricultural products. Extension services and digital platforms can also provide farmers with real-time guidance on best practices for post-harvest management.

Technology adoption is a game-changer in reducing post-harvest losses. Simple solutions, such as solar dryers for preserving fruits and vegetables or low-cost moisture meters for monitoring grain quality, can have a significant impact. Digital platforms that connect farmers with buyers in real time help reduce delays and ensure timely offtake of produce. Additionally, innovations like blockchain technology can enhance traceability, ensuring that produce meets quality standards throughout the supply chain.

Policy reforms are crucial for creating an enabling environment for post-harvest loss reduction. Governments can incentivize investments in storage and transportation infrastructure through tax breaks and subsidies. Establishing quality standards and regulatory frameworks for handling, storage, and transportation ensures accountability among supply chain actors. Additionally, integrating post-harvest loss reduction into national agricultural development plans highlights its importance and mobilizes resources for targeted interventions.

Public-private partnerships (PPPs) offer a powerful mechanism for addressing the systemic challenges associated with post-harvest losses in Nigeria. These collaborations combine the strengths of the public sector; policy direction, infrastructure investment, and regulatory oversight with the private sector's innovation, efficiency, and market reach. By working together, these stakeholders can design and implement solutions that are scalable, impactful, and sustainable.

One area where PPPs can have a significant impact is in the development of storage and logistics infrastructure. For instance, private companies can manage and operate cold storage facilities and silos built with public funding, ensuring their efficient use and maintenance. Governments can incentivize these partnerships by offering tax breaks, grants, or long-term leases for storage infrastructure in key agricultural zones. These facilities can also be integrated into agro-industrial parks, serving as central hubs for aggregation, storage, and processing.

Cold chain logistics is another area ripe for PPP innovation. Investments in refrigerated transport, temperature-controlled warehouses, and last-mile delivery systems can reduce losses for perishable goods like fruits, vegetables, dairy, and fish. Governments can provide co-financing or risk-sharing mechanisms to encourage private companies to enter this capital-intensive segment. For example, an initiative to deploy solar-powered refrigerated trucks across major agricultural corridors could drastically reduce spoilage while also supporting renewable energy goals.

Technology deployment is another area where PPPs can thrive. Collaborations between technology firms, agribusinesses, and government agencies can accelerate the adoption of digital platforms and innovative tools. For example, a partnership might develop an app that connects farmers with storage facilities and buyers in real time, streamlining the supply chain and ensuring timely offloading of perishable produce. Similarly, partnerships could fund pilot projects for advanced storage technologies, such as vacuum-sealed chambers or controlled atmosphere storage, which extend the shelf life of high-value crops.

PPPs can also play a critical role in financing and scaling farmer training programs. By pooling resources, public and private entities can develop and deliver workshops, field demonstrations, and digital tutorials on post-harvest handling and best practices. For example, a cocoa processing company might partner with local governments to train farmers on drying and fermenting techniques that preserve bean quality and reduce losses during storage.

Community-driven initiatives are an often-overlooked but vital component of reducing post-harvest losses. These grassroots efforts leverage the collective power of local actors, ensuring that solutions are tailored to the unique challenges and opportunities of specific regions. By empowering communities, stakeholders can create sustainable systems for managing agricultural produce post-harvest.

Farmer cooperatives are at the forefront of community-led solutions. These groups allow farmers to pool their resources, share knowledge, and access infrastructure that would be unaffordable individually. For example, cooperatives can collectively invest in drying equipment, grain silos, or cold storage units, ensuring that all members benefit from reduced losses and better market access. Cooperatives also provide a platform for collective bargaining, enabling farmers to secure fairer prices for their produce and reinvest in their operations.

Local governments and NGOs play a critical role in facilitating these community-led initiatives. They can provide seed funding, technical support, and training to help establish and sustain cooperatives. For example, an NGO might introduce solar drying technology to a farming community, train members on its use, and provide initial equipment on a cost-sharing basis. Over time, the cooperative could take ownership of the system, ensuring its long-term sustainability.

Educational campaigns within communities are another critical strategy. Farmers and supply chain actors often lack access to information about post-harvest management techniques. Community workshops, demonstration farms, and peer-to-peer learning networks can address this knowledge gap, equipping farmers with skills to reduce losses and maintain quality. For instance, a community initiative might train tomato farmers to use stackable plastic crates instead of traditional baskets to minimize bruising during transport.

Community-led initiatives also support the development of localized value chains. By focusing on processing and packaging agricultural products within the community, these efforts reduce the reliance on distant markets and create additional income streams. For example, a community with surplus cassava production might establish a small processing facility to produce and package cassava flour, which is less perishable and has higher market value.

9.1 Integrating Innovations for Long-Term Post-Harvest Loss Reduction

Innovation is critical to achieving long-term reductions in post-harvest losses in Nigeria. By embedding advanced technologies and scalable solutions into agricultural supply chains, stakeholders can enhance efficiency, minimize waste, and ensure that agricultural produce retains its value from farm to market. These innovations range from low-cost, farmer-friendly tools to cutting-edge technologies that revolutionize storage, transportation, and processing.

Low-cost innovations are particularly impactful for smallholder farmers, who often operate on tight budgets. Solar dryers, for instance, offer an effective solution for preserving fruits, vegetables, and grains by reducing moisture content and preventing spoilage. Unlike traditional drying methods that rely on open sunlight and are prone to contamination and inefficiency, solar dryers provide a controlled environment that maintains product quality. Similarly, hermetic storage bags designed to prevent pest infestations and fungal growth in grains are affordable, reusable, and easy to implement.

At a more advanced level, technologies such as IoT (Internet of Things) sensors and data analytics can transform how supply chains monitor and manage produce. IoT sensors placed in storage units or transport vehicles can provide real-time data on temperature, humidity, and gas levels, alerting operators to conditions that might compromise product quality. These sensors can also be integrated into digital platforms that provide actionable insights, enabling farmers, processors, and transporters to optimize their operations.

Artificial intelligence (AI) and machine learning are powerful tools for predicting and preventing post-harvest losses. For example, AI-driven systems can analyze weather patterns, market trends, and crop data to recommend the best harvesting and storage strategies. Machine learning algorithms can also predict spoilage risks based on historical data, helping supply chain actors make timely interventions. These technologies are particularly valuable for managing high-value perishable products like dairy, meat, and horticultural crops.

Blockchain technology offers a robust solution for enhancing transparency and accountability in the supply chain. By creating a digital ledger that records every transaction and movement of produce, blockchain ensures traceability and compliance with quality standards. For instance, exporters can use blockchain to verify that their goods meet international requirements, enhancing buyer confidence and reducing the risk of rejection at destination markets.

Collaborations with technology firms and research institutions are essential for integrating these innovations into Nigeria's agricultural supply chains. Pilot programs that test and refine technologies in local contexts can accelerate adoption and demonstrate their value to stakeholders. Moreover, partnerships between governments, private companies, and international donors can mobilize resources for scaling these solutions, ensuring that even smallholder farmers can access and benefit from advanced tools.

9.2 Building Resilience Against Future Challenges

Reducing post-harvest losses is not only about addressing current inefficiencies but also about building resilience against future challenges. Climate change, population growth, and evolving market demands will continue to place pressure on Nigeria's agricultural supply chains. Developing adaptive strategies and robust systems is essential to ensure that these supply chains remain efficient, sustainable, and responsive.

Climate change is a particularly pressing challenge, as rising temperatures, erratic rainfall, and extreme weather events disrupt farming and supply chain activities. Investments in climate-resilient infrastructure, such as flood-resistant storage facilities and drought-tolerant crop varieties, are crucial for mitigating these risks. For instance, farmers in flood-prone areas can benefit from raised storage units that protect produce during heavy rains, while regions experiencing prolonged dry spells can adopt water-efficient irrigation systems.

Population growth and urbanization are driving increased demand for food, requiring supply chains to scale up while maintaining efficiency. Addressing this demand necessitates the development of regional aggregation centers and processing hubs that can handle larger volumes of produce without compromising quality. These centers can also facilitate the integration of smallholder farmers into formal supply chains, providing them with access to larger markets and reducing the likelihood of produce being wasted.

Global market dynamics are another factor shaping the future of agricultural supply chains. As consumer preferences shift toward sustainable and ethically sourced products, Nigerian agricultural exports must align with these trends to remain competitive. This includes adopting certifications such as Global GAP, organic, and Fair Trade, which ensure compliance with international standards. By integrating these certifications into supply chains, Nigeria can enhance its reputation in global markets while promoting sustainable and equitable practices.

Resilience also requires strong governance and institutional support. Governments must prioritize investments in research and development, focusing on innovations that address post-harvest challenges. Additionally, regulatory frameworks must be strengthened to enforce quality standards, ensure fair trade practices, and provide incentives for stakeholders to adopt sustainable solutions. For example, establishing a national fund dedicated to reducing post-harvest losses can mobilize resources for critical interventions.

CHAPTER 10

Strategic Roadmap for the Future of Nigeria's Agricultural Supply Chains

The transformation of Nigeria's agricultural supply chains is critical for driving economic growth, enhancing food security, and positioning the country as a global agricultural powerhouse. The insights and strategies discussed throughout this book provide a foundation for developing a sustainable, inclusive, and efficient agricultural sector. To achieve this vision, Nigeria must focus on several key pillars: infrastructure development, technology and innovation, policy and governance, financing and investment, and capacity building and stakeholder engagement. These interconnected areas form the backbone of a strategic roadmap that can unlock the full potential of Nigeria's agriculture.

Infrastructure is the cornerstone of efficient agricultural supply chains. Addressing Nigeria's infrastructure deficits requires coordinated investments in transportation, storage, and processing facilities. Upgrading rural roads and expanding rail networks can improve connectivity between farms and markets, reducing transportation costs and minimizing losses during transit. Cold storage facilities, silos, and aggregation centers can preserve the quality of perishable goods and extend their shelf life, enabling farmers to access distant markets. Modern agro-industrial parks equipped with shared infrastructure for processing and packaging can further enhance value addition and competitiveness.

Technology and innovation are transformative forces that can revolutionize Nigeria's agricultural supply chains. Digital platforms connecting farmers with buyers and logistics providers eliminate inefficiencies and ensure timely transactions. Blockchain technology enhances transparency and traceability, particularly for export commodities that require compliance with stringent international standards. Precision agriculture tools, such as drones, soil sensors, and satellite imaging, enable farmers to optimize input use and improve productivity. Renewable energy solutions, such as solar-powered cold storage units and irrigation systems, align with sustainability goals while reducing operational costs.

Policy reform and effective governance are essential for creating an enabling environment for agricultural transformation. Land tenure systems must be streamlined to provide farmers with secure and transferable rights, encouraging long-term investments in productivity. Trade policies should align with regional agreements, such as the African Continental Free Trade Area (AfCFTA), to enhance competitiveness and simplify export procedures. Subsidies and incentives for adopting climate-smart technologies and practices can drive sustainability, while regulatory frameworks should enforce environmental and quality standards to ensure accountability.

Financing and investment are critical for addressing the financial challenges faced by farmers and supply chain actors. Expanding access to credit through innovative models, such as credit guarantee schemes and mobile banking solutions, can empower smallholder farmers to invest in their operations. Public-private partnerships (PPPs) can mobilize resources for infrastructure projects, such as cold chain logistics and agro-industrial parks. Additionally, targeted financing for value addition and export-oriented enterprises can enhance Nigeria's presence in global markets and boost foreign exchange earnings.

Empowering stakeholders with knowledge, skills, and resources is vital for sustained growth and innovation. Capacity-building programs can train farmers and supply chain actors in post-harvest management, sustainable practices, and market integration. Targeting women and youth with tailored support ensures equitable participation and leverages the potential of these often-overlooked groups. Strengthening cooperatives and farmer associations enhances collective bargaining power, while regular forums for stakeholder dialogue foster collaboration and alignment on shared goals.

Achieving the vision of a transformed agricultural sector in Nigeria demands a deliberate and sustained commitment to implementing the outlined strategies. The roadmap is not just a theoretical framework but a call to action for all stakeholders—farmers, policymakers, private sector actors, and development partners. The challenges are multifaceted but addressing them cohesively can unlock immense opportunities that extend beyond the agricultural sector, impacting the broader economy and social structure of the country.

Infrastructure investments must be prioritized to bridge the critical gaps in Nigeria's agricultural supply chains. For example, improving rural road networks not only facilitates the efficient movement of goods but also reduces isolation for farming communities, enabling them to access essential services and markets. The expansion of rail networks and inland waterway systems can significantly cut transportation costs for bulky commodities, creating a more integrated and cost-effective logistics framework. Additionally, the development of strategically located cold storage facilities and silos is essential for reducing post-harvest losses, particularly for high-value perishable goods like tomatoes, peppers, and dairy products.

Technology and innovation play a transformative role in addressing the systemic inefficiencies in Nigeria's agricultural supply chains. Precision agriculture tools, such as soil sensors and drones, can optimize resource use while boosting productivity. Blockchain technology, when integrated into supply chains, enhances traceability and compliance with international standards, opening up premium markets for Nigerian exports. Digital platforms that connect farmers with buyers, transporters, and input providers reduce transaction costs and improve market access. Furthermore, the adoption of renewable energy solutions, such as solar-powered irrigation and processing facilities, aligns with global sustainability trends while addressing local energy deficits.

Reforming policy frameworks and governance structures is equally critical. Simplifying land tenure systems to provide farmers with secure and transferable land rights will encourage long-term investments and innovation in farming practices. Harmonizing trade policies with regional agreements like the AfCFTA can create new market opportunities for Nigerian agricultural products, boosting the country's competitiveness on the global stage. Subsidies and incentives for adopting sustainable farming practices and technologies can accelerate the transition to a more resilient agricultural sector. Strong regulatory frameworks that enforce environmental, social, and quality standards ensure that agricultural development aligns with broader national and global goals.

Financing remains a significant bottleneck for many stakeholders in the supply chain, particularly smallholder farmers and small- and medium-sized enterprises (SMEs). Expanding access to affordable credit through innovative models, such as mobile banking and credit guarantee schemes, can empower farmers to invest in productivity-enhancing inputs and technologies. Public-private partnerships (PPPs) are indispensable for mobilizing resources for large-scale projects, such as agro-industrial parks and cold chain logistics. Targeted financing for value addition, such as processing facilities and export-oriented enterprises, can further enhance the profitability and sustainability of Nigeria's agricultural sector.

Capacity building and stakeholder engagement form the foundation of a sustainable transformation. Farmers must be equipped with the skills and knowledge to adopt modern practices, manage risks, and navigate complex markets. Tailored support for women and youth can address systemic inequities and unlock their potential as drivers of innovation and growth in the sector. Strengthening cooperatives and producer associations empowers farmers by enhancing their collective bargaining power and access to shared resources. Regular forums for stakeholder dialogue foster collaboration, ensuring that policies and initiatives reflect the needs and priorities of all actors in the supply chain.

The vision of a transformed agricultural supply chain in Nigeria is ambitious but attainable. It requires a unified effort to align investments, innovations, and policies with the goal of creating a resilient, sustainable, and inclusive agricultural sector. By committing to this roadmap, Nigeria can not only address its immediate challenges but also position itself as a leader in global agriculture. The benefits of this transformation extend far beyond increased productivity or export revenues—they include improved livelihoods for millions of Nigerians, enhanced food security, and a more diversified and robust economy.

The journey ahead will require perseverance, innovation, and collaboration across all levels. While the challenges are significant, the opportunities are even greater. Nigeria has the resources, talent, and potential to become a global agricultural powerhouse. By taking decisive action today, stakeholders can build a thriving agricultural sector that meets the demands of the future while fostering prosperity and sustainability for generations to come.

10.1 Driving Economic Growth Through Agricultural Industrialization

A reimagined agricultural supply chain has the potential to be a significant driver of Nigeria's economic growth. By transitioning from subsistence farming to industrialized agriculture, the country can create a multiplier effect that benefits not just the agricultural sector but also related industries, such as manufacturing, logistics, and technology. Agricultural industrialization involves increasing mechanization, fostering value addition, and building integrated systems that reduce inefficiencies and maximize productivity.

Mechanization is a crucial step toward industrializing Nigeria's agriculture. Most smallholder farmers still rely on manual labor and rudimentary tools, limiting their productivity and scalability. Access to modern machinery, such as tractors, harvesters, and irrigation systems, can drastically improve yields and reduce labor intensity. Governments and private investors can facilitate this transition through equipment leasing programs and subsidies for agricultural machinery. Mechanized farming not only increases efficiency but also opens up new opportunities for rural employment in machinery operation and maintenance.

Value addition through processing is another essential component of agricultural industrialization. Instead of exporting raw commodities,

Nigeria can focus on producing finished or semi-finished goods that fetch higher prices in both domestic and international markets. For instance, cashew nuts can be processed into roasted snacks or cashew milk, while cassava can be turned into starch, flour, or ethanol. Establishing processing hubs in agricultural regions ensures that value addition occurs closer to the source, reducing transportation costs and creating local jobs.

Industrialization also requires building integrated supply chains that connect production, processing, storage, and distribution seamlessly. This involves investing in infrastructure such as agro-industrial parks, cold chains, and logistics hubs. These systems ensure that agricultural products maintain their quality as they move through the supply chain, enhancing their marketability. Integration also fosters collaboration among stakeholders, from farmers and processors to distributors and retailers, creating a more efficient and cohesive ecosystem.

Ultimately, agricultural industrialization strengthens Nigeria's position in global markets. By producing high-quality, value-added products that meet international standards, the country can compete effectively in export markets, attract foreign investment, and increase its foreign exchange earnings. This transformation lays the groundwork for sustainable economic growth and diversification, reducing dependence on oil and creating a more resilient economy.

10.2 Empowering Communities for Sustainable Agricultural Development

At the heart of a transformed agricultural sector lies the empowerment of local communities. Nigeria's agricultural supply chains are heavily dependent on the efforts of smallholder farmers and rural communities, who face significant challenges, including limited resources, access to markets, and vulnerability to climate change. Empowering these communities is essential for ensuring the inclusivity and sustainability of agricultural transformation.

Education and training are foundational to community empowerment. Farmers must be equipped with knowledge about modern agricultural practices, post-harvest handling, and market dynamics. Extension services can serve as a vital bridge between research institutions and farmers, disseminating knowledge about climate-smart practices, pest control, and crop diversification. Digital tools and mobile apps can complement traditional extension services, providing farmers with real-time advice and access to online training resources. For example, an app that offers weather forecasts and planting recommendations can help farmers optimize their production cycles.

Financial inclusion is another critical aspect of empowerment. Many farmers lack access to formal credit, savings, and insurance, limiting their ability to invest in productivity-enhancing technologies. Expanding access to microfinance and mobile banking solutions ensures that farmers can secure the resources they need. Additionally, innovative financing models, such as crowdfunding platforms and value chain financing, can mobilize capital directly from investors and buyers, creating a more inclusive financial ecosystem.

Social inclusion must also be prioritized. Women, who make up a significant portion of Nigeria's agricultural workforce, often face systemic barriers to accessing land, credit, and decision-making opportunities. Programs that target women with tailored financing, training, and leadership opportunities can break these barriers and unlock their potential. Similarly, engaging youth in agriculture through mentorship programs, agribusiness incubation, and technology training ensures that the sector benefits from their energy and innovation.

Finally, fostering community ownership of agricultural initiatives ensures long-term sustainability. Cooperatives and producer organizations provide platforms for collective action, enabling farmers to pool resources, access shared infrastructure, and advocate for their interests. By empowering communities, Nigeria can build a resilient agricultural sector that not only drives economic growth but also improves livelihoods and fosters social equity.

REVIEWS

A Must-Read for Every Agricultural Stakeholder
By Adewale Abiodun, Agribusiness Consultant

Jumoke Ayodele Raji-Ayoola has delivered a masterpiece with *Agricultural Supply Chains in Nigeria: From Farm to Market Efficiency.* This book is a game-changer for anyone working in agriculture, from smallholder farmers to policymakers. The author's insights into the systemic bottlenecks in Nigeria's supply chains are eye-opening, and her solutions are practical and actionable. I particularly appreciate the emphasis on sustainability and inclusivity, which are often overlooked in discussions about agricultural transformation. This book has the potential to spark a revolution in Nigeria's agricultural sector. It's not just a book; it's a blueprint for progress.

A Comprehensive Guide to Transforming Nigerian Agriculture
By Ngozi Okafor, Agricultural Policy Analyst

This is one of the most insightful books I've read on Nigerian agriculture. Jumoke Ayodele Raji-Ayoola masterfully examines every aspect of agricultural supply chains, from production to market access, while proposing solutions that are grounded in reality. Her ability to connect macro-level policies with micro-level challenges faced by farmers is commendable. The book's focus on technology, infrastructure, and financing resonates deeply with the current needs of the sector. Whether you're a policymaker, a researcher, or simply passionate about agriculture, this book is an invaluable resource. Kudos to the author for producing such timely and relevant work.

A Visionary and Practical Approach
By Ibrahim Musa, Smallholder Farmer and Cooperative Leader

As a farmer, I found this book incredibly relatable and inspiring. Jumoke Ayodele Raji-Ayoola doesn't just talk about the problems; she provides clear, practical solutions that can be implemented at both the local and national levels. Her insights on empowering smallholder farmers, especially through cooperatives and value addition, hit close to home. The book also addresses issues like post-harvest losses and market access, which are major challenges for farmers like me. I would recommend this book to every farmer and agricultural leader—it's a guide we've been waiting for.

Transformative and Thought-Provoking
By Funke Adeyemi, Development Practitioner

This book is a testament to Jumoke Ayodele Raji-Ayoola's deep understanding of the agricultural sector and her commitment to creating a better future for Nigeria. The emphasis on sustainability and the integration of technology into agricultural supply chains are particularly impressive. Her discussions on policy reform are not just theoretical, they are actionable and tailored to Nigeria's unique context. This book is an excellent resource for development practitioners like me, as it provides the tools and insights needed to design effective interventions in agriculture. It's inspiring, practical, and essential reading for anyone passionate about Nigeria's growth.

Essential Reading for Policymakers and Entrepreneurs
By Tunde Olorunfemi, CEO of AgriConnect Nigeria

Jumoke Ayodele Raji-Ayoola has written what I believe is one of the most comprehensive books on Nigerian agriculture. As someone who works in agribusiness, I found her analysis of supply chain inefficiencies and her proposed solutions to be incredibly valuable. The focus on infrastructure, financing, and technology aligns perfectly with what the sector needs to thrive. I also appreciate the book's inclusivity, especially the attention given to empowering women and youth in agriculture. This is not just a book; it's a roadmap for building a sustainable and competitive agricultural sector. Every policymaker and entrepreneur in Nigeria should have this on their shelf.

ABOUT THE AUTHOR

Jumoke Ayodele Raji-Ayoola is a renowned agricultural economist, policy advisor, and development practitioner with over two decades of experience in transforming agricultural systems across Nigeria and Africa. Passionate about fostering sustainable and inclusive growth in the agricultural sector, she has dedicated her career to addressing systemic inefficiencies, empowering smallholder farmers, and advocating for innovative solutions to drive economic development.

Jumoke holds advanced degrees in Agricultural Economics and Rural Development from prestigious institutions, and her expertise spans agricultural supply chain management, policy reform, rural development, and agribusiness innovation. She has worked with numerous government agencies, international development organizations, and private-sector firms, contributing to groundbreaking initiatives that have improved livelihoods, enhanced food security, and boosted Nigeria's agricultural competitiveness.

An advocate for sustainability and inclusivity, Jumoke has been a vocal supporter of women and youth empowerment in agriculture. Through her leadership, she has championed programs that provide access to finance, training, and markets for marginalized groups, ensuring that agriculture becomes a tool for equitable development.

Jumoke is also a prolific writer and speaker, known for her ability to translate complex agricultural concepts into practical strategies that resonate with diverse audiences. Her research and publications have been widely recognized, and she frequently speaks at international conferences and forums on the future of agriculture in Africa.

When she's not working on agricultural initiatives, Jumoke is deeply involved in mentoring young professionals and entrepreneurs in the agricultural space. She believes that the future of agriculture lies in the hands of innovative and passionate individuals who are committed to creating sustainable solutions.

Agricultural Supply Chains in Nigeria: From Farm to Market Efficiency reflects Jumoke's deep knowledge, experience, and vision for Nigeria's agricultural sector. Through this book, she provides a comprehensive roadmap for transforming supply chains and unlocking the immense potential of agriculture to drive economic growth, sustainability, and resilience.